A User's Guide to
ELLIPSOMETRY

HARLAND G. TOMPKINS

D1453016

DOVER PUBLICATIONS, INC.
MINEOLA, NEW YORK

Bibliographical Note

This Dover edition, first published in 2006, is an unabridged
republication of the work originally published in 1993 by Academic
Press, Inc., San Diego, CA.

International Standard Book Number: 0-486-45028-7

Manufactured in the United States of America
Dover Publications, Inc., 31 East 2nd Street, Mineola, N.Y. 11501

Contents

Preface

Ellipsometry is a very old technique for studying surfaces and thin films. It is also used extensively in integrated circuit (IC) manufacture and development, in corrosion science, and various other areas. The fundamental principles of the technique are well known. Although the literature is extensive, the few available books deal with the subject on a level that is too mathematical or theoretical for most casual users. Most users are skilled in microelectronics IC fabrication, corrosion science, etc., and are not particularly skilled in physical optics.

Many commercial instruments are available and most present-day instruments utilize an associated microcomputer of some form. Software is normally available so that for certain specified tasks, the system can be used in a turnkey manner without knowing very much about how the ellipsometer does what it does. However, the tasks which can be done in a turnkey manner represent only a small subset of the possible applications of the technique.

This book is intended for the technologist who has an instrument available and who wants to use it in its present form without significant modification. We will assume that the instrument is kept in working condition by the instrument manufacturer's technical representative. Correspondingly, we will not spend any time on alignment of instruments or on instrument design aspects. The angle of incidence on most turnkey instruments is 70° and a He-Ne laser with a wavelength of 6328 Å is used as the light source. These values will be generally used in this text unless stated otherwise.

This book is specifically designed for the technologist who wishes to stretch the use of the technique beyond the turnkey applications. We shall deal with situations where all of the critical information is not available and those where approximations are required.

Ellipsometry is an optical technique. In order for the user to understand how the ellipsometer does what it does, we shall begin by providing some of the fundamentals of optics that are particularly applicable. We take these from physics textbooks on optics, taking only those ideas applicable to our purposes here. These will lead us to the definitions of Del and Psi, which are the parameters an ellipsometer measures. We shall refer to the change in values of Del and Psi as a function of film thickness, index of refraction, etc., as the "Del/Psi trajectories." Understanding these trajectories is key to understanding the technique and stretching it beyond the turnkey applications. We have tried to keep the theoretical aspects short and to the point, including only those ideas that are directly related to the subject at hand.

As indicated above, this book is intended for users who already have an operating ellipsometer. It consists of seven chapters followed by 14 case studies. Chapter 2, on "Instrumentation" briefly describes how the commercially available instruments work. A chapter entitled "Using Optical Parameters to Determine Material Properties" deals with how one goes from the measured optical parameters to deducing information about the material being studied. The chapter on "Determining Optical Parameters for Inaccessible Substrates and Unknown Films," shows what to do when all of the necessary information is not available.

A question often asked is "How thin a film can be measured with ellipsometry?" Chapter 5 considers this question and uses some examples of the adsorption of single layers of atoms. In the microelectronics industry, a commonly used material is "polysilicon." Chapter 6 deals with the difficulties specific to this material when using ellipsometry. Some suggestions are made for dealing with some of the difficulties. Finally, the last chapter discusses the effect of substrate roughness.

Examples are the best way to show how ellipsometry can be stretched beyond the turnkey applications. Fourteen case studies are included to illustrate various uses of ellipsometry. These are works of others taken from the reviewed literature.

Finally, we include an appendix on calculating the Del/Psi "trajectories" discussed in the text, an appendix on "Effective Medium Considerations" in which we discuss how one calculates an "effective" index of refraction with mixtures of materials or rough surfaces, and an appendix where we list some of the literature values of the optical constants of various materials.

No one writes a book such as this alone, without significant input from others. Correspondingly, I would like to acknowledge the management at Bell Laboratories and at Motorola for allowing me the necessary time for the exploration needed to understand and use this method and to acknowledge my co-workers at these institutions for stimulating me to do so. In addition, I would like to acknowledge my wife, Rose Ann Tompkins, for encouraging me to write the book and for spending many hours alone without complaint while I sat in front of a computer, buried in the intricacies of composing the book.

Chapter 1
Theoretical Aspects

Electromagnetic waves and polarized light are treated in textbooks[1-5] and reference books[6] on optics. We review here some of the salient features that are directly applicable to ellipsometry.

1.1 Description of an Electromagnetic Wave

The electromagnetic wave is a transverse wave consisting of both an electric field vector and a magnetic field vector, which are mutually perpendicular and are perpendicular to the propagation direction of the wave. It can be specified with either the magnetic field vector or the electric field vector. For simplicity, we shall consider the electric vector only. The light wave can be represented mathematically as

$$A = A_o \sin\left(-\frac{2\pi}{\lambda}(x - vt) + \xi\right) \qquad (1)$$

where A is the electric field strength of the wave at any given time or place. A_o is the maximum field strength and is called the "amplitude," x is the distance along the direction of travel, t is time, v is the velocity of the light, λ is the wavelength, and ξ is an arbitrary phase angle, which will allow us to offset one wave from another when we begin combining waves.

If we consider the wave at a fixed time, the variation of the electric field with position can be represented as shown in Figure 1-1. We can identify locations where the electric field is maximum, minimum, and zero. In this particular case, the electric field variation occurs only in the vertical direction.

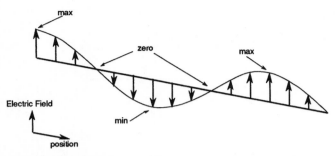

Figure 1-1. An electromagnetic wave at a fixed time, represented schematically.

Waves transport energy, and the amount of energy per second that flows across a unit area perpendicular to the direction of travel is called the "intensity" of the wave. It can be shown[7] that the energy density is proportional to the square of the amplitude and that the intensity I is given by

$$I = cA^2/8\pi \qquad (2)$$

where c is the speed of light.

1.2   Interaction of Light with Material

1.2.1 The Complex Index of Refraction

Suppose we have light passing from one medium (e.g., ordinary room air) into another medium that is not totally transparent, as suggested by Figure 1-2. Several phenomena occur when the light passes the interface. One phenomenon that occurs is some of the light is reflected back and does not enter the second medium. We shall deal with this reflected component later. For the moment, let us consider the light that enters the second medium.

The parameter we use to describe the interaction of light with the material is the complex index of refraction Ñ, which is a combination of a real part and an imaginary part and is given as

$$\tilde{N} = n - jk \qquad . \qquad (3)$$

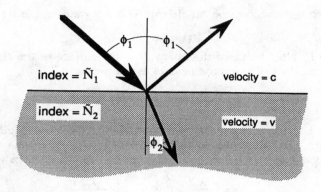

Figure 1-2. Light reflecting from and passing through an interface between air and a material characterized by the complex index of refraction $\tilde{N}_2$.

"n" is also called the "index of refraction" (this sometimes leads to confusion), and "k" is called the "extinction coefficient". (In the scientific literature, the imaginary number, the square root of -1, is denoted either as i or j. We will use j.) For a dielectric material such as glass, none of the light is absorbed and k = 0. In this case, we are only concerned with n.

The velocity of light in free space is usually designated as "c" and its value is approximately $c = 3 \times 10^{10}$ cm/sec. The velocity of light in air is not significantly different from this value. When light enters another medium, however, the velocity will be different. Let us designate its value in the medium as "v." We define the index of refraction "n" to be

$$n = c/v. \tag{4}$$

Clearly, the value of n for free space is unity.

Before we define the extinction coefficient "k," let us consider first the absorption coefficient "α." In an absorbing medium, the decrease in intensity I per unit length z is proportional to the value of I. In equation form this is

$$\frac{dI(z)}{dz} = -\alpha\, I(z) \tag{5}$$

where α is the absorption coefficient. This integrates out to

$$I(z) = I_o e^{-\alpha z} \tag{6}$$

where I_o is the value of the intensity at the surface of the absorbing medium. The extinction coefficient k is defined as

$$k = \frac{\lambda}{4\pi}\alpha \quad \text{or} \quad k = \frac{n\lambda_m}{4\pi}\alpha \tag{7}$$

where λ is the vacuum wavelength and λ_m is the wavelength in the medium. From Equation 6, the value of I(z) is 1/e (approximately 37%) of the value of I_o when $\alpha z = 1$. This occurs when $z = 1/\alpha$ or when

$$z = \frac{\lambda}{4\pi k} \quad \text{or} \quad z = \frac{\lambda_m}{4\pi}\frac{n}{k} \quad . \tag{8}$$

1.2.2 Laws of Reflection and Refraction

When the light beam reaches the surface, as suggested in Figure 1-2, some of the light is reflected and some passes into the material. The law of reflection says that the angle of incidence is equal to the angle of reflection, i.e.,

$$\phi_i = \phi_r \quad . \tag{9}$$

In the figure, both are shown as ϕ_1. The part of the light beam that enters the material at an angle ϕ_1 does not continue in the same direction, but is refracted to a different angle ϕ_2. The law of refraction is called "Snell's law" after its discoverer (in the early 1600s) and is given by

$$\tilde{N}_1 \sin \phi_1 = \tilde{N}_2 \sin \phi_2 \tag{10}$$

for materials in general. For dielectrics, k = 0 and Equation 10 becomes

$$n_1 \sin \phi_1 = n_2 \sin \phi_2 \quad . \tag{11}$$

Equation 11 consists of only real numbers, and hence is reasonably straightforward. In Equation 10, k for the first medium usually is zero, hence $\tilde{N}_1 = n_1$. If $\tilde{N}_2$, on the other hand, is complex, i.e., k_2 is nonzero, then the angle ϕ_2 is complex and the quantity no longer has the simple significance of the angle of refraction.[8]

1.2.3 Dispersion

Up to this point, we have referred to the real and imaginary parts of the index of refraction as "n" and "k". It should be mentioned that these are not simple constants for a given medium, but are in fact functions of the wavelength λ. This is why when white light enters a prism, it emerges with the various colors separated.

We use the term "dispersion" to describe how the optical constants change with wavelength. The equation for $n(\lambda)$ is often approximated by

$$n\,(\lambda) = n_1 + \frac{n_2}{\lambda^2} + \frac{n_3}{\lambda^4} \tag{12}$$

where n_1, n_2, and n_3 are called the "Cauchy coefficients." The equation for $k(\lambda)$ is approximated by

$$k\,(\lambda) = k_1 + \frac{k_2}{\lambda^2} + \frac{k_3}{\lambda^4} \tag{13}$$

where k_1, k_2, and k_3 are called the "Cauchy extinction coefficients." Figures 1-3 and 1-4 show typical examples of the variation of n and k as a function of wavelength.

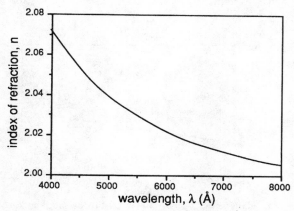

Figure 1-3. Index of refraction n as a function of wavelength for silicon nitride.[9]

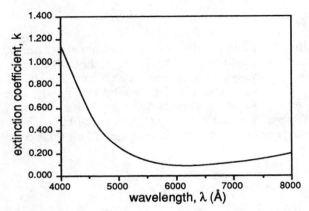

Figure 1-4. Extinction coefficient k as a function of wavelength for polycrystalline silicon.[10]

1.2.4 The Effect of Temperature

The optical constants are functions of temperature, in addition to being functions of wavelength. Figure 1-5 shows an example of how n varies with temperature. Figure 1-6 shows corresponding data for silicon. Small variations in temperature are insignificant, but for *in situ* experiments where the material may be at elevated temperatures, one must take into account this variation.

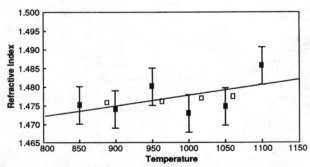

Figure 1-5. The value of the refractive index of silicon dioxide as a function of temperature in °C. (After Yu[11])

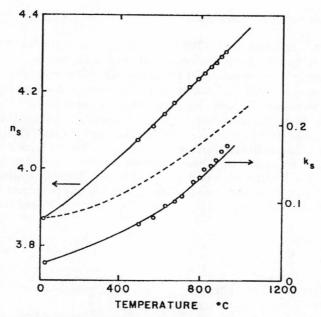

Figure 1-6. Temperature dependence of the refractive index of silicon. Circles are determined ellipsometrically in present experiments, the dashed curve is that given by Ibrahim and Bashara.[12] (After Hopper[13])

1.3 Polarized Light

Most light sources emit light that has components with electric fields oriented in all of the possible directions perpendicular to the direction of travel. We refer to this as unpolarized light. If all of the photons in a light beam have the electric field oriented in one direction, the light is referred to as polarized or, more completely, linearly polarized light. Some light sources emit polarized light. In addition one can obtain polarized light by passing the light beam through an optical element or by causing the beam to make a reflection under some specific conditions.

1.3.1 Linearly Polarized Light

Let us suppose that we have two light beams with the same frequency moving along the same path, one polarized in the vertical plane and the other polarized perpendicular to the vertical plane, as suggested in Figure 1-7. For simplicity, let us assume that the amplitude of both waves are the same. Suppose that the maxima of the two beams coincide. This is the same as saying that the phase is the same. These two beams can be combined to give a resultant light beam that is also linearly polarized. In our illustration, we have suggested that the intensity of the two beams is the same and hence the resultant beam would be linearly polarized at 45° from each of the original beams. If the intensities were not the same, the beams would still have been linearly polarized, but the angle of polarization would have been different from 45°. The key point

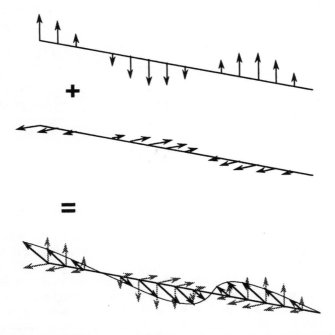

Figure 1-7. If two linearly polarized light beams which are in phase are combined, the resultant light beam is linearly polarized.

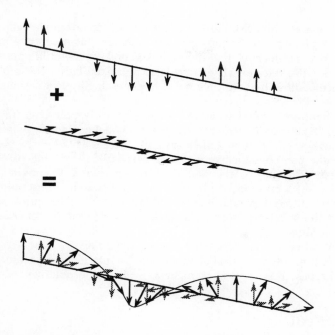

Figure 1-8. If two linearly polarized light beams which are out of phase are combined, the resultant light beam is elliptically polarized. In this particular example, they are out of phase by 90°, hence the resultant beam is circularly polarized.

here is that **when two linearly polarized waves** with the same wavelength (or frequency) **are combined in phase, the resultant wave is linearly polarized.**

1.3.2 Elliptically Polarized Light

Suppose we have two light beams traveling along the same path and that they have the same wavelength. As opposed to the previous example, let us suppose that the maxima do not coincide, but are out of phase by some amount (in Figure 1-8, the maximum of one coincides with the zero of the other, giving a phase difference of 90°). When these two waves are combined, the tips of the arrows do not move back and forth in a plane as in the previous example. In this case, they move around in a manner such that if you looked end on, it would appear to be a circle. We refer to this as circularly polarized light. If the phase difference is other than 90°, we have, in general, elliptically polarized light.

There are several ways to obtain elliptically polarized light in practice. Of primary interest to us is the fact that when linearly polarized light makes a reflection on a metal surface, there is a shift of the phases of both the components (parallel and perpendicular to the plane of incidence). The shift is, in general, not the same for both components, hence the resultant light will be elliptically polarized. The amount of ellipticity induced depends on various things including the optical properties of the substrate as well as thickness and optical properties of overlying films.

Another way to change the ellipticity of polarized light is to pass the beam through an optical element called a quarter-wave plate. We shall discuss this optical element as well as the polarizer later in Chapter 2.

1.4 Reflections

1.4.1 Coordinate System for Reflections

Ellipsometry invariably involves the reflection of light from a surface. To describe this reflection, let us suppose that we have an incident plane wave moving in the direction suggested in Figure 1-9 and suppose that this wave makes a reflection from the surface as suggested

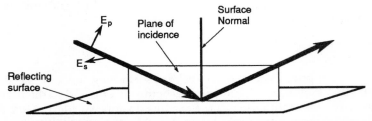

Figure 1-9. Reflection of a light beam from a surface. The plane of incidence contains both the incoming beam and the outgoing beam as well as the normal to the surface.

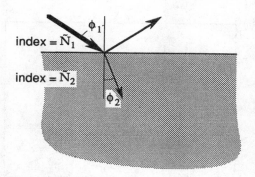

Figure 1-10. Reflection and transmission at a single interface.

in the figure. Let us define the plane of incidence as being that plane that contains the light beam prior to and after the reflection. The plane of incidence also contains the normal to the surface.

We shall refer to plane polarized waves that are in the plane of incidence as "p waves" and plane waves polarized perpendicular to the plane of incidence as "s waves" ("s" comes from the German word "senkrecht").

1.4.2 Fresnel Reflection Coefficients and the Brewster Angle

Suppose that we have a light beam making a reflection at an interface between medium 1 and medium 2 as shown in Figure 1-10. Some of the light is reflected and some is transmitted. The Fresnel reflection coefficient r is the ratio of the amplitude of the reflected wave to the amplitude of the incident wave for a single interface. The Fresnel reflection coefficients are given[1,6] by

$$r_{12}^p = \frac{\tilde{N}_2 \cos\phi_1 - \tilde{N}_1 \cos\phi_2}{\tilde{N}_2 \cos\phi_1 + \tilde{N}_1 \cos\phi_2} \qquad r_{12}^s = \frac{\tilde{N}_1 \cos\phi_1 - \tilde{N}_2 \cos\phi_2}{\tilde{N}_1 \cos\phi_1 + \tilde{N}_2 \cos\phi_2} \qquad (14)$$

where the superscript refers to waves parallel or perpendicular to the plane of incidence and the subscript refers to either medium 1 or medium 2. Corresponding equations exist for transmission, but these are not used in ellipsometry.

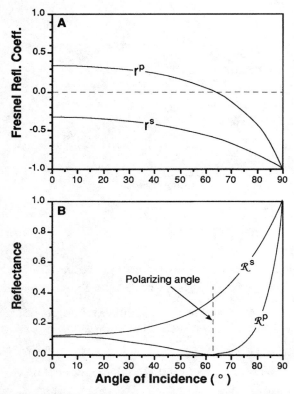

Figure 1-11. (A) The ratios of the amplitudes of the reflected wave to the amplitudes of the incident wave (the Fresnel coefficients) as a function of angle of incidence ϕ_1 for a material with index $\tilde{N}_2 = 2.0 - 0.0\,j$. (B) The ratios of the intensities of the reflected wave to the intensities of the incident wave (the reflectance) as a function of angle of incidence ϕ_1 for the same material. The polarizing angle is shown by the dashed line.

Recall that the intensity of the light is proportional to the square of the amplitude. We define the reflectance $\mathcal{R}$ as the ratio of the reflected intensity to the incident intensity. For a single interface, this can be written as

$$\mathcal{R}^P = |r^P|^2 \quad \text{and} \quad \mathcal{R}^s = |r^s|^2 \quad . \tag{15}$$

If both media are dielectric in nature (i.e., k = 0) then the complex index of refraction $\tilde{N}_i$ becomes simply n_i. It is useful to illustrate some of the features of these equations with an example. Let us suppose that medium 1 is air, with $n_1 = 1.0$, and medium 2 is a dielectric such as silicon nitride ($n_2 = 2.0$). Figure 1-11A shows a plot of the Fresnel coefficients (Equation 14) as a function of the angle of incidence. Figure 1-11B shows the resulting reflectances $\mathcal{R}^p$ and $\mathcal{R}^s$ from Equation 15.

At normal incidence, $\phi_1 = 0$ and from Snell's law, $\phi_2 = 0$ and all of the cosines in equation 14 are equal to +1. We have, then,

$$r_{12}^p = \frac{n_2 - 1}{n_2 + 1} \qquad r_{12}^s = \frac{1 - n_2}{1 + n_2} \qquad \mathcal{R}^p = \mathcal{R}^s = \left(\frac{n_2 - 1}{n_2 + 1}\right)^2 . \quad (16)$$

For the example mentioned, the reflectance for both the parallel and perpendicular component is about 11%. Note that at normal incidence the distinction between parallel and perpendicular disappears.

For other than normal incidence, we see from Figure 1-11A that r^s is always negative and nonzero. This follows from the fact that for $0 < \phi < 90°$, $\cos \phi$ is always positive and hence the denominator of both equations is always positive and nonzero. In our example, we have chosen $n_2 > n_1$. One of the results of this from Snell's law, Equation 10, is that $\phi_2 < \phi_1$ and hence $\cos \phi_2 > \cos \phi_1$. From this we can see that $n_2 \cos \phi_2 > n_1 \cos \phi_1$, which implies that the numerator of the equation for the s wave is always negative and nonzero. Hence $\mathcal{R}^s$ is always positive and nonzero.

Figure 1-11A shows that r^p changes from positive to negative, i.e., goes through zero, as the angle of incidence increases. At this point, the reflectance $\mathcal{R}^p$ is also zero. This too can be shown mathematically. Let us consider the numerator for the p wave. From Snell's law,

$$n_2 \cos \phi_2 = \sqrt{n_2^2 - n_1^2 \sin^2 \phi_1} . \quad (17)$$

We can use this equation to eliminate ϕ_2 from the numerator and express it in terms of ϕ_1, n_2, and n_1. For the specific angle ϕ_1 such that

$$\tan \phi_1 = \frac{n_2}{n_1} , \quad (18)$$

it can be readily shown by algebraic manipulation that the numerator is zero.

The significance of this is that when unpolarized light makes a reflection with this specific angle of incidence, none of the light polarized parallel to the plane of incidence is reflected, i.e., it is all transmitted. The resulting reflected light is polarized perpendicular to the plane of incidence. This specific angle given by Equation 18 is called various names, including the "polarizing angle," the "Brewster angle," and the "principal angle." For the example shown in Figure 1-11, with $n_2 = 2.0$, the Brewster angle is about 63.43°. This is indicated by the dashed line.

This is the principle by which polarized sunglasses remove the glare from the highway or water surface in front of the user. Most of the light polarized parallel to the plane of incidence has been removed by the reflection. Glare is caused by the remaining reflected light polarized perpendicular to the plane of incidence. The sunglasses are simply polarizers, which remove most of this component. Light reflected from a water surface will be more intense than light transmitted from below the surface. This glare often inhibits a person above the surface from viewing features (fish or alligators) below the water's surface. A photographer may use a polarizer to remove this component thus allowing the light transmitted from below the surface to be used for a good photo.

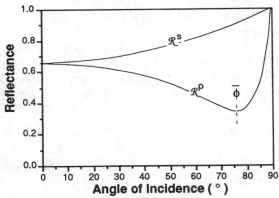

Figure 1-12. The ratios of the intensities of the reflected wave to the intensities of the incident wave (the reflectance) as a function of angle of incidence ϕ_1 for a material with index $\tilde{N}_s = 2.01 - 3.75\,j$ (e.g., for Ni). The principal angle is indicated by the dashed line.

It might be noted that although all of the reflected light is polarized in one direction, the transmitted light still has components from both polarizations, with the amount of perpendicularly polarized light reduced. Since about 36% of the perpendicularly polarized light was reflected, the transmitted beam contains only 64% of the incident perpendicularly polarized light.

Multiple interfaces can be used to remove successively more and more of the perpendicularly polarized light until the transmitted light is virtually pure polarized light in the plane of incidence. Historically, this is one method of obtaining polarized light.[14]

When the reflecting surface is not a dielectric, i.e., k is nonzero, the situation becomes more complicated. The Fresnel reflection coefficients are, in general, complex numbers, so the concept of being greater than zero, passing through zero, and being less than zero is no longer appropriate. Because of this, there is no plot analogous to Figure 1-11A for metals. The reflectance, however, is a real number. Figure 1-12 shows the reflectance for both the p wave and the s wave for Ni. This is similar to the corresponding plot for dielectrics, except that the reflectance does not go to zero. The angle for which $\mathcal{R}^p$ is minimum is called the principal angle of incidence and is denoted by $\overline{\phi}$. At the principal angle of incidence the phase difference between the two components is 90°.

It might be noted that high reflectance is obtained when the index of the substrate is significantly different from that of the ambient. The numerator of the Fresnel coefficients will be larger under these conditions. This will occur when either n_s is significantly different from unity or when k_s is large.

1.4.3 Total Reflection Coefficients for Multiple Interfaces

The above discussion concerned a single interface. We simply disregarded the light transmitted through the interface. Many real world situations involve multiple interfaces as suggested by Figure 1-13.

The resultant reflected wave returning to medium 1 is made up of the light reflected directly from the first interface plus all of the transmissions from the light approaching the first interface from medium 2. Each successive transmission back into medium 1 is smaller. The addition of this infinite series of partial waves leads to the resultant wave. Both Azzam[15] and Heavens[16] derive the ratio of the amplitude of the resultant

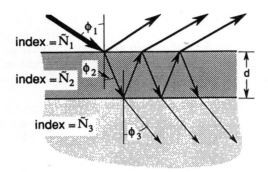

Figure 1-13. Reflections and transmissions with multiple interfaces.

reflected wave to the amplitude of the incident wave and it is given by the total reflection coefficients

$$R^P = \frac{r_{12}^P + r_{23}^P \exp(-j\,2\beta)}{1 + r_{12}^P\,r_{23}^P \exp(-j\,2\beta)} \qquad R^s = \frac{r_{12}^s + r_{23}^s \exp(-j\,2\beta)}{1 + r_{12}^s\,r_{23}^s \exp(-j\,2\beta)} \quad (19)$$

where the subscript "23" denotes that this Fresnel reflection coefficient is for the interface between medium 2 and medium 3. β is the film phase thickness and is given by

$$\beta = 2\pi \left(\frac{d}{\lambda}\right) \tilde{N}_2 \cos\phi_2 \qquad . \qquad (20)$$

where d is the film thickness. By algebraic manipulation, it can be shown that as $d \to 0$ then the total reflection coefficients become equal to a Fresnel coefficient between medium 1 and medium 3.

When multiple interfaces are present, the reflectance $\mathcal{R}$ is given by

$$\mathcal{R}^P = |R^P|^2 \quad \text{and} \quad \mathcal{R}^s = |R^s|^2 \qquad (21)$$

rather than by Equation 15.

1.5 Ellipsometry Definitions

1.5.1 Del and Psi

Referring to Figure 1-9, let us denote the phase difference between the parallel component and the perpendicular component of the incoming wave as δ_1. Further, let us denote the phase difference between the parallel component and the perpendicular component of the outgoing wave as δ_2. Let us define the parameter Δ, called delta or often abbreviated "Del" as

$$\Delta = \delta_1 - \delta_2 . \tag{22}$$

Del, then, is the change in phase difference that occurs upon reflection and its value can be from zero to 360°.

Without regard to phase, the amplitude of both perpendicular and parallel components may change upon reflection. $|R^p|$ and $|R^s|$ from Equation 19 are the ratios of the outgoing wave amplitude to the incoming wave amplitude for the parallel and perpendicular components, respectively. We define the quantity Ψ in such a manner that

$$\tan \Psi = \frac{|R^p|}{|R^s|} . \tag{23}$$

Ψ then is the angle whose tangent is the ratio of the magnitudes of the total reflection coefficients. The value of Ψ can be between zero and 90°.

1.5.2 The Fundamental Equation of Ellipsometry

We define the complex quantity ρ (rho) to be the complex ratio of the total reflection coefficients such that

$$\rho = \frac{R^p}{R^s} . \tag{24}$$

The fundamental equation of ellipsometry[17] then is

$$\rho = \tan \Psi \, e^{j\Delta} \qquad \text{or} \qquad \tan \Psi \, e^{j\Delta} = \frac{R^p}{R^s} \tag{25}$$

where Δ and Ψ are the quantities measured by an ellipsometer. The information about the sample in question is contained in the total reflection coefficients, or R^p and R^s. It should be noted that assuming the instrument is operating properly, the quantities Δ and Ψ which are measured are correct. Whether the calculated sample parameters such as thickness and index of refraction are correct depend on whether the model assumed is correct. Suppose one assumes that the sample is a film-free substrate when in fact the sample is a thin film of one material on top of a substrate of another. The measured quantities Δ and Ψ will be correct, but the calculated quantities n and k for the sample will be incorrect.

1.6 References

1. "Fundamentals of Optics," F. A. Jenkins and H. E. White, 3rd Edition, McGraw-Hill, New York (1957).
2. "Geometrical and Physical Optics," J. Morgan, McGraw-Hill, New York (1953).
3. "Optics," F. W. Sears, Addison-Wesley, Reading, MA (1958).
4. "Optics," M. V. Klein, John Wiley, New York (1970).
5. "Light," G. S. Monk, Dover, New York (1963).
6. "Principles of Optics," M. Born and E. Wolf, 4th Edition, Pergamon Press, New York (1969).
7. "Fundamentals of Optics," F. A. Jenkins and H. E. White, 3rd Edition, McGraw-Hill, New York (1957), p. 414.
8. "Principles of Optics," M. Born and E. Wolf, 4th Edition, Pergamon Press, New York (1969), p. 615.
9. W. A. Pliskin, *J. Electrochem. Soc.*, **134**, 2819 (1987).
10. Values of the Cauchy extinction coefficients for polysilicon provided by Mark Keefer, Prometrix Corporation, Santa Clara, CA 95054.
11. C. T. Yu, K. H. Isaak, and R. E. Sheets, *Semiconductor International*, May 1991, p. 166.
12. M. M. Ibrahim and N. M. Bashara, *J. Vac. Sci. Technol.*, **9**, 1259 (1972).
13. M. A. Hopper, R. A. Clarke, and L. Young, *Surface Sci.*, **56**, 472 (1976).
14. "Polarisation of Light," W. Spottiswoode, Macmillan & Co., London (1874).
15. "Ellipsometry and Polarized Light," R. M. A. Azzam and N. M. Bashara, North-Holland Publishing Co., Amsterdam (1977), p. 283.
16. "Optical Properties of Thin Solid Films," O. S. Heavens, Dover Publications, New York (1965), p. 55.
17. "Ellipsometry and Polarized Light," R. M. A. Azzam and N. M. Bashara, North-Holland Publishing Co., Amsterdam (1977), p. 287.

Chapter 2
Instrumentation

2.1 Fundamentals and History

The instrumentation of ellipsometry as described herein requires the following:

1. A monochromatic light source.
2. An optical element to convert unpolarized light to linearly polarized light.
3. An optical element that converts linearly polarized light into elliptically polarized light.
4. A reflection from the sample of interest.
5. An optical element to determine the state of polarization of the resultant light beam.
6. A detector to measure the light intensity (or to determine the presence of a null).
7. Calculation facilities to interpret the results in terms of an assumed model of the sample.

It should be noted that although it was not called by that name, ellipsometry was practiced[1] in the late 1800s. Monochromatic light sources have been available for a long time in the form of sodium or mercury arc lamps. These sources give out several discrete wavelengths. All but one are filtered out. Optical elements such as polarizers and quarter-wave plates have also been available for some time. In a small treatise[2] on polarization of light published in 1874 , Spottiswoode describes how polarized light can be obtained by passing light through a Nicol prism and by reflection. He also describes a "quarter undulation plate."   In the list above, items # 2 and #5 are both basically polarizers.

Spottiswoode distinguishes between #2, the polarizer which first *causes the light to be polarized* (which he calls the "polariser") and #5, the polarizer which is used to *determine the state of polarization* (which he calls the "analyser"). One might recall that in 1874, light was still being considered a motion of the "ether" that was understood to pervade all space.

The name "ellipsometry" was introduced in 1945 in an article[3] published in *Review of Scientific Instruments* by Alexandre Rothen entitled "The Ellipsometer, an Apparatus to Measure Thicknesses of Thin Films." For many years, the detector used to determine the null position of the analyzer was the human eye.

Figure 2-1 shows an instrument illustrated in a book by Drude,[4] published in 1901. Although Drude did not call this instrument an ellipsometer, it has a polarizer p, an analyzer p', a compensator C, and the surface in question S. The light source would be brought to the telescope F and the detector (the eye) would be brought to the telescope K.

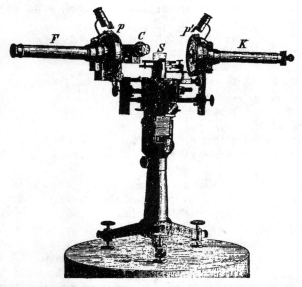

Figure 2-1. Early optical instrument used for studying changes in polarization upon reflection. Telescopes F and K, polarizers p and p', compensator C, and reflecting surface S. (After Drude[4])

2.2 Optical Elements

2.2.1 The Light Source

In general, we will assume that our light source is unpolarized. Any given electromagnetic wave leaving a light source will be polarized in some direction. Unpolarized light results from the superposition of many individual contributions which are randomly oriented. Monochromatic light is obtained by using a laser, by using an arc lamp, or by using a polychromatic source and filtering. Historically, the arc lamps were used. Today, the source typically will be a laser. In some instances where two or three specific wavelengths are needed, a polychromatic source with filters will be used.

2.2.2 Polarizers and Analyzers

In this section, we discuss the function of the polarizer and quarter-wave plate. We do not discuss how these optical elements do what they do. Suppose, as suggested in Figure 2-2, that we have a light wave that is polarized at an angle of 45° from the vertical. The polarizer is an optical element which has a polarizing axis as suggested in the figure by the heavy dotted line. If the axis of the polarizer lines up with the angle of polarization of the wave, the entire wave is transmitted, as shown in Figure 2-2A. When the axis does not line up with the angle of polarization of the wave, we separate the wave into components which are parallel to and perpendicular to the polarizer axis, as suggested in Figures 2-2B and 2-2C. The component that aligns with the axis is transmitted and the component that is perpendicular is blocked. If the axis of the polarizer is perpendicular to the angle of polarization of the wave, as suggested in Figure 2-2D, no light is transmitted and we have a "null."

Polarizers are used in two different ways. If the polarizer is used to convert unpolarized light to polarized light, it is called the "polarizer." If it is used to determine the state of polarized light by locating the null, it is called the "analyzer." This terminology dates from the early years of studying polarized light and was used by Spottiswoode.[2] For reflection ellipsometry, the angle position of the polarizer is the angle between the polarizing axis and the plane of incidence, measured clockwise from the plane of incidence as the observers looks in the direction of propagation. Typically, both a polarizer and an analyzer are used in ellipsom-

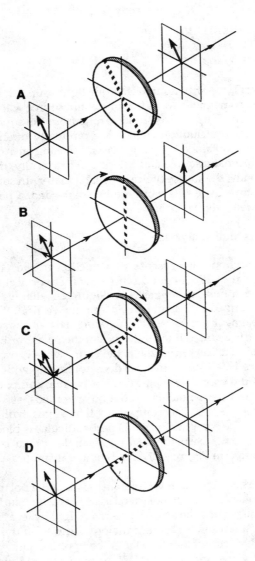

Figure 2-2. Polarizer with its axis (shown as the heavy dashed line) oriented at various angles to the polarization direction of the incident light beam. The polarizer lets through only that component of the light which is aligned with its axis.

etry. Their angles are usually denoted by "P" and "A," respectively. Note that any given polarizer position specified by the angle α is redundant with the angle $\alpha + 180°$. If the polarizer is aligned with the polarized light, moving the polarizer through an angle of 90° will cause the amount of transmitted light to change from all of it to none of it.

2.2.3 The Quarter-Wave Plate

The next optical element to be considered is called by various names including the "retarder," the "compensator," or the quarter-wave plate. Wave plates in general are anisotropic optical elements. The velocity of the wave depends on its orientation. In general, the wave plate has a fast axis and a slow axis, both of which are perpendicular to each other and to the direction of propagation of the wave. The component of the wave which is aligned with the fast axis passes through the optical element faster than the component aligned with the slow axis. If the two components of the wave were in phase before passing through the element (i.e., linearly polarized) then, in general, they will be out of phase when they emerge (elliptically polarized). The thickness of the wave plate can be chosen such that the phase difference is exactly 90°, 180°, or 360° and they are called respectively quarter-wave plates, half-wave plates, or full-wave plates. In ellipsometry, we use the quarter wave-plate (which we will denote as the "QWP"). Note that this element is a quarter-wave plate for only one wavelength.

We illustrate the quarter-wave plate in Figure 2-3 where the dark line indicates the fast axis and the lighter line indicates the slow axis (perpendicular to the fast axis). Figure 2-3A suggests that when the fast axis is horizontal, the horizontal component of the wave emerges from the QWP 90° ahead of the component which was aligned with the slow axis. Figure 2-3B suggests that when the original wave (before being broken into its components) is aligned with the fast axis, both components are retarded the same amount and hence the wave emerges still in phase. Figure 2-3C indicates that when the fast axis is vertical, the horizontal component of the light is retarded by a quarter wave compared to the vertical component. Figure 2-3D indicates that when the original wave is aligned with the slow axis, both components are retarded the same amount and again, as when aligned with the fast axis, the two components emerge in phase.

If we use the vertical component as the reference, we would describe the retardation as going from -90° in Figure 2-3A through zero

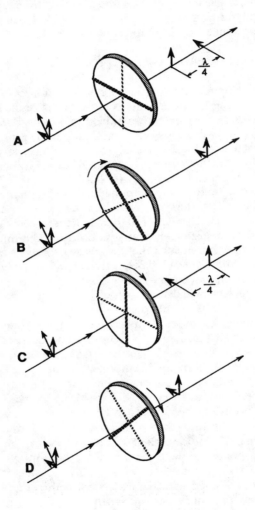

Figure 2-3. The quarter-wave plate. The fast axis is indicated by the dark line and the slow axis is indicated by the lighter line. The component aligned with the fast axis emerges a quarter wave ahead of the component aligned with the slow axis.

in Figure 2-3B to +90° in Figure 2-3C back to zero in Figure 2-3D. If the two components were equal in amplitude, this would represent going from circularly polarized to linearly polarized to circularly polarized in the opposite direction and then back to linearly polarized.

2.2.4 The Reflection

In general, the reflection can induce a phase shift from zero to 360° and can attenuate either or both of the components. If a wave with both components equal (polarized at 45° from the plane of incidence) makes a reflection from a surface and neither were attenuated, the resultant wave would appear end-on as shown in Figure 2-4A for various phase shifts. With attenuation where the p wave is attenuated more than the s wave, the resultant wave would appear as shown in Figure 2-4B.

2.2.5 The Detector

Whereas historically, the eye was used to determine when the null was present, photomultipliers are universally used today. Some manual ellipsometers require the use of the eye to set the polarizer and analyzer near null, after which the photomultiplier is used for the precision measurement.

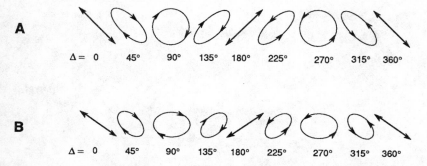

Figure 2-4. If the incident wave has equal components which are in phase, the resultant wave from a reflection is shown by (A) if no attenuation (or the same attenuation) were present for various phase shifts, and (B) if the p wave were attenuated more than the s wave.

2.3 The Manual Null Instrument

Before getting into the automated instruments, we will describe the manual null instrument so as to understand the function of each component.

2.3.1 Conceptual

Prior to discussing the actual instruments used, we discuss the concepts involved. The first requirement is to have monochromatic polarized light. We suggest in Figure 2-5 that we use a laser that is passed through a polarizer. If we were to then make a reflection on the surface, the light would, in principal, come out elliptically polarized as shown in Figure 2-4. Linearly polarized light is much simpler to analyze since we can use another polarizer (called the analyzer) to determine the null and hence the angle of polarization. In general, however, the light will not be linearly polarized after the reflection, but will be elliptically polarized, as indicated by Figure 2-5. We need to change the light from elliptically polarized light into linearly polarized light and we use the QWP to do this.

Let us digress for a moment and consider elliptically polarized light as suggested by Figure 2-6. In Figure 2-6A we have two perpendicular components (with the coordinate system in the vertical and

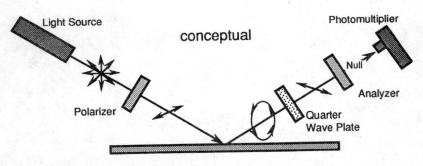

Figure 2-5. If linearly polarized light makes a reflection, the resultant light is elliptically polarized. Conceptually, the QWP could change it back to linearly polarized light and the analyzer could find a null. The angular settings of the QWP and the analyzer could be used to determine the phase shift and attenuation ratio. This is a conceptual use of these elements. Actual practice is somewhat different.

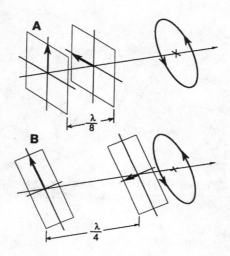

Figure 2-6. For elliptically polarized light, the phase difference between the components depends on which coordinate system is being used.

horizontal directions), which are equal in amplitude, and 1/8 of a wavelength out of phase. In Figure 2-6B we have the same elliptically polarized light. In this case, the coordinate system for the components is rotated 45° from that used in Figure 2-6A. The elliptically polarized light is made of components that have unequal amplitude and are 1/4 wave out of phase. Clearly, the value of the phase difference between the components of any given elliptically polarized light depends on the coordinate system being used for the components.

Any elliptically polarized light can be converted into linearly polarized light by placing a quarter-wave plate in the path and aligning the fast and slow axes with the appropriate axes of the ellipse. This is depicted in Figure 2-7.

Returning to Figure 2-5, we use the QWP to return the elliptically polarized light to linearly polarized light. When the QWP is at the correct angle, the analyzer will be able to find a null. If the QWP is not at the correct angle, the analyzer will not be able to extinguish the light. The angles of the QWP and the analyzer required to do this provides information as to the phase difference and attenuation caused by the reflection and hence allow us to calculate Del and Psi.

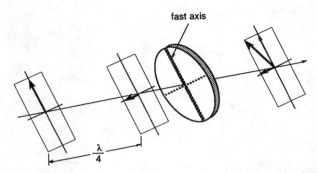

Figure 2-7. Elliptically polarized light can be changed into linearly polarized light by aligning the axes of a QWP with the appropriate axes of the ellipse.

2.3.2 Actual Practice

Figure 2-5 is useful to explain conceptually how the QWP and analyzer can determine the phase difference and orientation. In common practice, however, the QWP is placed before the reflection as shown in Figure 2-8. Rather than holding the polarizer fixed and rotating the QWP and analyzer, the QWP is set at an angle of 45° and the polarizer and analyzer are rotated until the null is found. One of the reasons for using this configuration is that the relationship between the settings of the polarizer and the analyzer and the values of Del and Psi is simpler.

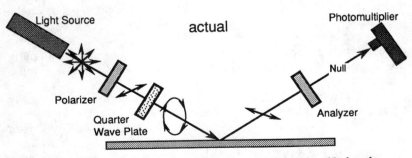

Figure 2-8. In actual practice of null ellipsometry, the QWP is placed before the reflection and is held in a fixed position. The polarizer and analyzer are rotated to find the null.

The polarizer-QWP combination functions as an elliptical polarizer. The ellipticity is adjusted so that it is exactly canceled by the reflection.

The angular convention is that all angles for the polarizer, analyzer, and QWP are measured as positive counterclockwise from the plane of incidence when looking into the beam. Del and Psi are a result of the reflection, and we use the polarizer, QWP, and analyzer to determine the values. By using the polarizer and analyzer in the above manner, we can obtain several combinations that will give the null. By removing redundancies (angles 180° from each other), by restricting the angle of the QWP to one angle (e.g., 45°), and by restricting the ranges of the polarizer and analyzer, the various combinations can be reduced to two zones,[5] defined as

Zone 2: $-45° < P_2 < 135°$, $0° < A_2 < 90°$, $QWP = 45°$.   (1)
Zone 4: $-135° < P_4 < 45°$, $-90° < A_4 < 0°$, $QWP = 45°$.   (2)

The relationship of Del and Psi to P and A determined in these two zones is

$Del_2 = 270° - P_2$, $Del_4 = 90° - 2P_4$
$Psi_2 = A_2$, $Psi_4 = -A_4$.   (3)

If the compensation of the QWP is exact, one can use either zone to determine Del and Psi. Using the average can eliminate errors due to inexact compensation. In fact, the difference between the values can be used to determine the compensation of the QWP.[6]

Figure 2-9 shows a plot of the intensity of the light near a null. One should note that very near the null, the change in intensity as a function of angle is very small. In addition, near the null, the light level

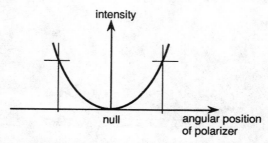

Figure 2-9. The null can be located more accurately if angular measurements are made on both sides of the null at equal intensity. The null is the average angular value.

is very low and noise considerations can sometimes be a problem. Since the curve is symmetric about the null, another way to find the null is to read the angle on both sides of the null at equal intensities and then take the average. This will give a more accurate value of the angular position of the null than reading it directly.

2.4 Rotating Element Instruments

2.4.1 The Rotating Null Instrument

The instruments already described have elements which are manually rotated by the operator. For many instruments in use today, however, under microprocessor control, the instrument rotates its own elements as part of its operation. There are basically two different types of rotating element instruments. One of these is a null instrument and is illustrated in Figure 2-10. When the analyzer is rotated about the light beam, the photomultiplier will detect a sinusoidally varying intensity. In general, the minima do not correspond to extinction of the light (null) and the intensity does not go to zero. When the polarizer is at the proper orientation so that the ellipticity is just canceled by the reflection, the light being detected will be linearly polarized and the maxima will be greatest and the minima will be zero (null). The proper orientation is

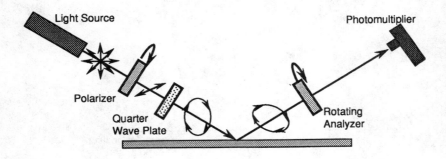

Figure 2-10. One type of rotating element ellipsometer. This type is a null instrument. The polarizer and analyzer rotate alternately until the null is found.

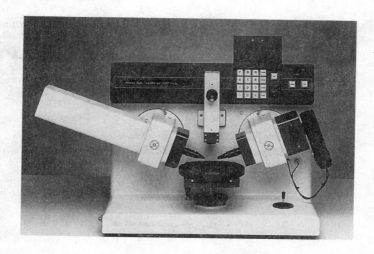

Figure 2-11. One example of a rotating null ellipsometer. (Rudolph Research, One Rudolph Rd., Flanders, NJ 07836)

found by adjusting alternately the polarizer and the analyzer to deepen the minima until true extinction is found.[7] Although this type of instrument uses rotating elements, it is still a null instrument, using the rotating elements to find the null.

A microprocessor is usually used to control the instrument and to do the calculations. The operator has a choice of using both null positions or only one. The first is more accurate, but the second is faster. In making a series of measurements, one can measure both null positions the first time and then use the information obtained to allow for more accurate single null measurements on subsequent samples. Figure 2-11 shows one example of a rotating null instrument.

2.4.2 The Photometric Instrument

The second type of rotating element instrument is a photometric instrument (it uses the values of the light intensity). This instrument is shown schematically in Figure 2-12. In this case, the polarizer is set at a

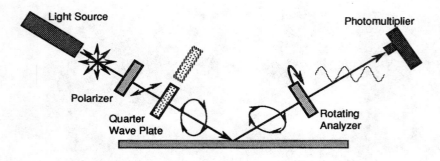

Figure 2-12. A second type of rotating element instrument. In this type, only the analyzer rotates. The intensity measurement can be made with the QWP either in the beam or out of the beam.

fixed value, usually 45°. The QWP is fixed with the fast axis perpendicular to the plane of incidence and can be placed in the beam or removed from it. Only one of the polarizing elements (the analyzer) is rotated. The intensity I measured as a function of time t by the photomultiplier can be expressed as

$$I(t) = I_0[1 + \alpha \cos 2A(t) + \beta \sin 2A(t)\ ] \qquad (4)$$

where

$$A(t) = 2\,\pi\,f\,t + A_c \qquad (5)$$

with f being the angular frequency of the rotation and A_c being a constant phase offset. I_0 is the average intensity and α and β are the Fourier coefficients. Aspnes has shown[8] that for a rotating analyzer instrument, the optimum precision is obtained when the light incident on the rotating analyzer is circularly polarized. If the phase shift induced by the reflection were about 90°, the optimum precision would be realized by using linearly polarized light (no QWP). The light starts out linearly polarized (components in phase) and the reflection changes it into near circularly polarized light. If, on the other hand, the phase shift due to the reflection was either none or about 180°, without the QWP, the light incident on the analyzer would be almost linearly polarized, or far from optimum. In this case, the QWP is used. Typically, Ψ is measured without the QWP in place and Δ is measured both with and without the

QWP. Depending on the value obtained, one or the other of the measurements is used.

The Fourier coefficients α and β are given[9] by

$$\alpha = -\cos 2\Psi \quad \text{and} \quad \beta = \sin 2\Psi \cos \Delta_m \tag{6}$$

where Δ_m is the measured phase difference between the parallel and perpendicular components of the light arriving at the analyzer. These equations can be inverted to give

$$\cos \Delta_m = \pm \sqrt{\frac{\beta^2}{1 - \alpha^2}} \quad \text{and} \quad \tan \Psi = \sqrt{\frac{1 + \alpha}{1 - \alpha}} \tag{7}$$

Figure 2-13. One example of a rotating element photometric ellipsometer. (Gaertner Scientific Corp., 1201 Wrightwood Ave., Chicago, IL 60614)

If the QWP was not used (i.e., before the reflection, the two components had zero phase difference), then the phase shift induced by the reflection is given by $\Delta = \Delta_m$. If the QWP was in place and the light incident on the sample was circularly polarized, then it is necessary to add the retardation of the QWP to the measured value. If the QWP is perfect, then $\Delta = \Delta_m + 90°$.

As indicated above, the instrument makes measurements with the QWP and without the QWP. Ψ is calculated from the measurement without the QWP. Making measurements in both configurations allows the resolving of the "±" sign for Δ as well as obtaining the most precise value. Figure 2-13 shows an example of a rotating element photometric instrument.

2.5 References

1.   P. Drude, *Wied. Ann.*, **43**, 126 (1891).
2.   "Polarisation of Light," W. Spottiswoode, Macmillan & Co., London (1874).
3.   A. Rothen, *Rev. Sci. Instruments*, **16**, 26 (1945).
4.   "Theory of Optics," P. Drude, Longmans, Green & Co., New York (1901), p. 258.
5.   "Semiconductor Material and Device Characterization," D. K. Schroder, Wiley-Interscience, New York (1990), p. 463.
6.   F. L. McCrackin, *Nat. Bur. Stand. Tech. Note 479* (1969).
7.   R. F. Spanier, *Industrial Research* (Sept. 1975).
8.   D. E. Aspnes, *J. Opt. Soc. Am.*, **64**, 639 (1974).
9.   P. S. Hauge and F. H. Dill, *IBM Journal of Research & Development*, **17**, 472 (1973).

Chapter 3
Using Optical Parameters to Determine Material Properties

The point was made in Chapter 1, and we repeat it here, that contrary to popular belief, ellipsometers **do not** measure the thickness of films. Ellipsometers measure the quantities Δ and Ψ. We use these quantities along with an assumed model to calculate material properties. The reliability of the calculated properties is only as good as the assumed model. If an improper model is assumed, although the values of Δ and Ψ are correct, the calculated quantities may well be meaningless.

In Chapter 2 we discussed how ellipsometers measure Δ and Ψ. In this chapter, we will discuss how the material quantities are calculated and will introduce the concept of Δ/Ψ trajectories.

We will be dealing with substrates and films. A material is considered to be a substrate if we do not have to deal with any lower boundaries or any material underneath it. Typically this would be because the value of k and the thickness are such that the light is essentially totally absorbed before it reaches the lower material boundary. In some instances we would consider a dielectric material to be a substrate if the lower boundary is far enough away or rough enough that light reflecting from the lower boundary does not return to the ellipsometer measuring system. A film, on the other hand, will be a material transparent enough that the underlying layer must be taken into account.

3.1 Del/Psi and n and k for Substrates

In most of our measurements, we are interested in a film on top of a substrate. To determine the properties of the film, we must know

35

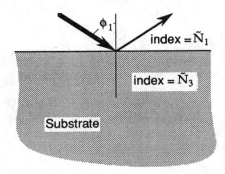

Figure 3-1. A light beam reflecting from a substrate, shown schematically.

something about the substrate. It is specifically necessary to know the optical constants $\tilde{N}_3 = n_3 - jk_3$ of the material. We use ellipsometry to determine these values of n_3 and k_3.

The subscript "1" will be used for the ambient and the subscript "3" will be used for the substrate, anticipating when we will later want to use the subscript "2" for a film between the substrate and the ambient. Suppose that we have a sample with a reflecting light beam as indicated in Figure 3-1. With our ellipsometer we measure Δ and Ψ. We can calculate the complex value of ρ from Equation 25, Chapter 1. The complex value of $\tilde{N}_3$ is then[1]

$$\tilde{N}_3 = \tilde{N}_1 \tan \phi_1 \sqrt{1 - \frac{4 \rho \sin^2 \phi_1}{(\rho + 1)^2}} \tag{1}$$

If we prefer using only real numbers, this can be separated into the equations[2]

$$n_3^2 - k_3^2 = n_1^2 \sin^2 \phi_1 \left[1 + \frac{\tan^2 \phi_1 \ (\cos^2 2\Psi - \sin^2 \Delta \sin^2 2\Psi)}{(1 + \sin 2\Psi \cos \Delta)^2} \right] \tag{2}$$

and

$$2 n_3 k_3 = \frac{n_1^2 \sin^2 \phi_1 \ \tan^2 \phi_1 \ \sin 4 \Psi \sin \Delta}{(1 + \sin 2\Psi \cos \Delta)^2} \tag{3}$$

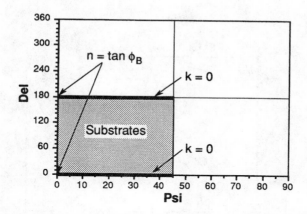

Figure 3-2. For substrates, Del will be 180° or 0 for dielectrics and between 0 and 180° for nondielectrics. Psi will fall between 0 and 45°.

The Del/Psi point for substrates falls in the lower left-hand quadrant, as suggested by Figure 3-2. Let us consider the above equations and some limiting values. Del = 0 or Del = 180 occurs when the substrate is a dielectric. From Chapter 1, we know that the phase shift from a reflection from a dielectric is either zero if the angle of incidence is less than the Brewster angle or 180° if it is more than the Brewster angle. A value of either zero or 180° will cause sin D to be zero. From Equation 3, this implies that either n_3 or k_3 will be zero and, of course, for dielectrics, k_3 is zero. Note that the Brewster angle will be 70° if $n_3 = 2.747$.

A value of Psi = 0 will cause the right-hand side of Equation 3 to be zero. For dielectrics, Psi = 0 means that the component of light in the plane of incidence is zero and this occurs when the angle of incidence is at the Brewster angle, where the phase shift is indeterminate between zero and 180°.

The right-hand side of Equation 3 being zero implies that either n_3 or k_3 is zero, but it does not tell us which is the case. If we insert Ψ = zero into Equation 2, we obtain a result for the right-hand side of Equation 2 that is always positive. Because of the minus sign in the left-hand side of Equation 2, we would have a contradictory result if n_3 were zero, hence this tells us that in general (not just for dielectrics) the Psi = 0 line corresponds to k_3 = zero. Since k_3 = zero implies that we have a

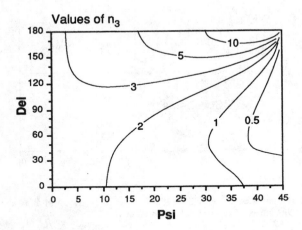

Figure 3-3. The loci of all Del/Psi points for substrates having the indicated values of n_3.

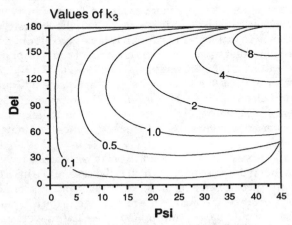

Figure 3-4. The loci of all Del/Psi points for substrates having the indicated values of k_3.

dielectric, and this implies that Del = zero or 180°, we conclude that the Psi = zero line is not a possibility for substrates.

Let us defer consideration of the value Psi = 45° for the moment. By using McCrackin's program,[3] we can determine the relationship between n_3/k_3 and Del/Psi (for n_1 = 1.0, ϕ_1 = 70°, and λ = 6328 Å). Figure 3-3 shows the loci of all Del/Psi points that have the indicated values of n_3. Similarly, Figure 3-4 shows the loci of all Del/Psi points that have the indicated values of k_3. Our conclusions above about k_3 = 0 for situations where Del = 0 or 180 and Psi = 0, and the conclusion about Psi = 0 being off-limits are substantiated by Figures 3-3 and 3-4.

Let us now consider the case of Psi = 45°. The right-hand side of Equation 3 becomes zero. This implies that either n_3 or k_3 is zero. By substituting Psi = 45° into Equation 2, we see that the right-hand side changes from negative to positive as Del passes from below approximately 40° to above 40°. Figures 3-3 and 3-4 suggest that if Del < 40°, then k_3 = 0. If Del > 40°, n_3 = 0. Since k_3 = 0 with Del being anything but 0 or 180° is not allowed, and since n_3 = 0 is unrealistic, we see that the line Psi = 45° is also off-limits for substrates.

Summarizing, then, for substrates, Del can be equal to 0 or 180° if k_3 = 0 and between these values for nonzero values of k_3. Psi can be greater than 0 and less than 45°.

Most ellipsometers have a software program to measure Del and Psi and then to calculate the substrate parameters n_3 and k_3. One might wonder why it is necessary to measure these optical parameters since it would seem that these kind of quantities would have been previously measured and tabulated in handbooks. Unless the material is extremely reproducible (such as single crystal silicon), the value of the optical constants depend on crystalline structure, grain size, etc. The value of n for polycrystalline silicon[4] can vary from about 3.85 to about 4.6 and the value of k can vary from about 0.02 to 0.30 depending on the deposition temperature. This emphasizes the importance of measuring the substrate parameters for the particular samples being studied.

3.2   The Calculation of Del/Psi Trajectories for Films on Substrates

Let us suppose that we have a substrate covered by a film as suggested in Figure 3-5. We shall call a film "transparent" if the extinction coefficient is zero (k_2 = 0). With given values of n_3, k_3, n_2, k_2, and

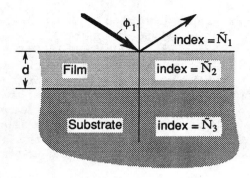

Figure 3-5. Schematic drawing of a film with index $\tilde{N}_2$ on top of a substrate with index $\tilde{N}_3$.

thickness, we can use Equations 10, 14, 19, 20, and 25 from Chapter 1 to calculate the expected values of Del and Psi. McCrackin[3] published a FORTRAN program in 1969 which makes various ellipsometry calculations. Among these is the calculation of a table of values of Del and Psi for various film thicknesses. We use a modified version of McCrackin's program to calculate these Del/Psi trajectories. Although various computer languages have been developed since 1969, FORTRAN is still the best language since it handles complex numbers well. The original program was written for batch processing with cards. Appendix A gives FORTRAN listings of programs similar to the table part of McCrackin's program, which has been modified for use on a Macintosh desktop computer. Most of the Del/Psi trajectories used in this book were calculated using these programs.

3.3 Trajectories for Transparent Films

Figure 3-6 illustrates the Del/Psi trajectory for silicon dioxide on silicon. The film-free value simply corresponds to the silicon substrate. In this case, $\Delta^\circ = 178.5^\circ$ and $\Psi^\circ = 10.5^\circ$, corresponding to $\tilde{N}_3 = 3.872 - j\,0.037$. If we begin to add a film with index $\tilde{N}_2 = 1.46$, the Del/Psi point begins to move down and to the right, tracing out the trajectory. The Del/Psi trajectory is indicated by the solid line and the values for thicknesses in 100 Å increments are shown as the squares.

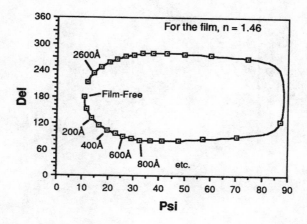

Figure 3-6. The Del/Psi trajectory for silicon dioxide on silicon with angle of incidence $\phi_1 = 70°$ and wavelength $\lambda = 6328$ Å.

An unknown thickness of a silicon dioxide on silicon sample can be determined by inspection by comparing the measured values of Del and Psi with this curve. For films with $k = 0$, the trajectory closes on itself at a thickness of

$$d = \frac{\lambda}{2 \sqrt{n_2^2 - \sin^2\phi_1}} \tag{4}$$

In this case the thickness d is about 2832 Å. For larger thicknesses, the trajectory simply retraces the same path. To determine thickness from a measured Del/Psi point, it is necessary to know, from some other source, the appropriate period of the the Del/Psi trajectory. One often knows roughly the thickness of the film from oxidation time, deposition time, etc.

On some microcomputer programs, the first several possible values are displayed. Note that the period depends on the wavelength and on the angle of incidence in addition to the index of the film. If one has the liberty to change either of these, there is a method of unambiguously

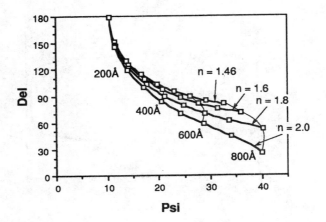

Figure 3-7. The Del/Psi trajectories for films with several different indices of refraction on single crystal silcon substrates. The first 800 Å is shown.

determining thickness. By changing either the wavelength or the angle of incidence, a different set of possible thickness values would be obtained. Only one value should be the same on both lists. This value would represent the correct thickness.

The trajectory shown in Figure 3-6 is for one particular index of refraction for the film. Other values of the index would give different trajectories. Figure 3-7 shows the trajectory for the first 800 Å for films with indices of 1.46, 1.6, 1.8, and 2.0. We note that in some regions, these curves are well separated. This can be used to determine the index of refraction of the film if it is not already known. The curve on which the measured point falls determines the index of refraction. The position on that curve determines the thickness.

One Del/Psi point gives us two numerical values. With this it is reasonable to expect to calculate two unknowns about the film. For any given film, we have three parameters, namely, n, k, and the thickness. If we assume that k = 0, it is reasonable to expect to be able to determine both the thickness and the value of n.

Figure 3-8 shows essentially the complete trajectories for films with indices of refraction of 1.46, 1.65, and 2.0 on single crystal silicon.

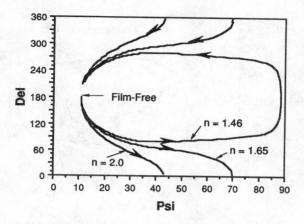

Figure 3-8. The Del/Psi trajectories for films having indices of refraction of 1.46, 1.65, and 2.0 on single crystal silicon. Approximately one period is shown.

These indices are typical of silicon dioxide, photoresist, and silicon nitride, respectively. Although two of these trajectories appear to be discontinuous, one should recall that for a fixed value of Ψ, the value of $\Delta = 0$ and $\Delta = 360$ are the same. We are attempting to plot what is essentially a hemisphere onto a flat piece of paper. Since all of these films are on the same substrate, they start at the same Del/Psi point. The thicknesses at which each trajectory returns to the film-free point (the period point) are 2832 Å, 2333 Å, and 1792 Å, respectively.

If one is using this method for determining the value of the index of refraction, we note that there are regions of the Del/Psi domain which provide much better information than others. Near the period point (or the film-free point) the curves are not particularly well separated. The Del/Psi trajectories cannot be used effectively in this region to determine the index of refraction. Sometimes, however, one has the flexibility to choose the thickness of the film being used for the index determination. In this situation, the best choice is the film whose thickness is not particularly close to the period point. We note that there is a "sweet spot" where the value of Ψ is the greatest.

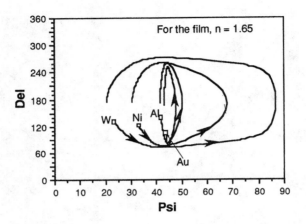

Figure 3-9. Del/Psi trajectories for an organic film with n = 1.65 on several metals. The indices for the substrates are given in Table 3-1.

TABLE 3-1. Optical Constants for Several Metals, along with Corresponding $\Delta°$, $\Psi°$, and Depth to 1/e.

Metal	n	k	$\Delta°$	$\Psi°$	Depth to 1/e
W	3.41	2.63	130.0°	23.3°	191 Å
Ni	2.01	3.75	121.0°	33.2°	134 Å
Al	1.21	6.92	140.4°	41.9°	73 Å
Au	0.16	3.21	104.7°	43.7°	157 Å
Si	3.85	0.02	178.5°	10.5°	2.5 μ

n & k values from *Amer. Inst. of Phys. Handbook*, D. E. Gray, Coordinator Editor, 3rd Edition (1972), McGraw Hill, New York

As a further example of films on other substrates, Figure 3-9 shows the trajectory of a hydrocarbon with index n = 1.65 on several metals. For illustration purposes, we have taken the index for the various substrates from a handbook and they are listed in Table 3-1. We emphasize that this is for illustrative purposes only; in actual practice, one should measure these for the particular samples of interest.

Whereas silicon dioxide on silicon uses most of the Del/Psi domain, an organic layer on aluminum or gold is rather crowded in the center. Since the period depends on the film only, the period for each of these is 2333 Å.

3.4 Trajectories for Absorbing Films

A film of material with nonzero extinction coefficient k can be considered an absorbing film or a substrate, depending on its thickness. Suppose that we are going to deposit a film of absorbing material onto a substrate gradually. When the thickness is zero, the Del/Psi point will be representative of the substrate material. When the thickness is very thick, the Del/Psi point will be representative of bulk film material. As we progress from no film to a very thin film to a thicker film, the Del/Psi trajectory will move from the value characteristic of the substrate to the value characteristic of bulk film material. We show this graphically in Figure 3-10.   In this case, we are depositing a film of tungsten onto

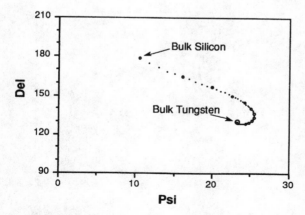

Figure 3-10. The Del/Psi trajectory when a film of tungsten is deposited onto silicon. The small dots are at 10 Å intervals. The large dots are at 50 Å intervals from zero thickness up to 500 Å.

silicon. We have used handbook values for the index of refraction of tungsten for illustrative purposes. We again caution the reader to determine the appropriate values for each particular sample.   The values of the index and the corresponding Del/Psi point for both material are given in Table 3-1. The small dots represent increments of thickness of tungsten of 10 Å and the heavier dots represent increments of 50 Å up to 500 Å. The Del/Psi point that represents bulk tungsten is shown as the larger open circle.

Initially, the points are well separated. As the thickness increases, however, the points become closer together until eventually, the separation between adjacent points is within the experimental measurement uncertainty. The trajectory begins to spiral into the target value. The fact that this is a spiral is not evident from Figure 3-10, but it will become evident later.

Table 3-1 gives the depth where the intensity has decreased to 1/e of its surface value as 191 Å for tungsten. During the first 191 Å, the points are well separated. During the second and third 191Å, they are much closer together and by the fourth 191 Å, the points are experimentally indistinguishable from the bulk tungsten point.

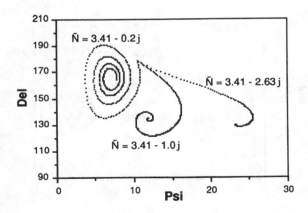

Figure 3-11. The trajectories when tungsten and two hypothetical materials with the indicated indices are deposited onto silicon. This shows the effect of k on the trajectory. Larger values of k cause the spiral to close more rapidly.

The value of k determines how fast the spiral closes on the bulk value. In Figure 3-11, we show the same curve as in Figure 3-10, where k = 2.63. In addition, we show the curve for two hypothetical materials which have the same value of n as tungsten, but different values of k. In the first case, we hypothesize a material with index $\tilde{N}$ = 3.41 - 1.0 j. The Del/Psi point for bulk material with this index is Del = 134.3° and Psi = 11.89°. The fact that k is smaller causes the spiral to close more slowly. To further illustrate this, we hypothesize a material with an even smaller value of k, which is more characteristic of a semiconductor than a metal. In this case, $\tilde{N}$ = 3.41 - 0.2 j, and the corresponding Del/Psi point is Del = 165.4° and Psi = 7.07°. Because k is much smaller, the spiral closes very slowly, with large maximum and minimum excursions. For even smaller values of k, the trajectory may pass through zero and eventually cover a large portion of the Del/Psi region. The limiting case, of course, is for k = 0 where the spiral never goes to the target value. In this case, the trajectory closes on itself and continues periodically, as described in the previous section.

3.5 Two-Film Structures

3.5.1 Transparent Films on Top

For two films, it is necessary to use a combination of the one-film program and a modification of the three-film program given in Appendix A. Let us first consider a dielectric film on top of another dielectric film. In Figure 3-12, we start with a film-free substrate of silicon, with the Del/Psi point at 178.5°/10.5°. Let us first suppose that we will have a nitride film being deposited on top of 1100 Å of silicon dioxide on top of the silicon substrate. For reference, we show the trajectory for silicon nitride only (also shown in Figure 3-8).

The trajectory for the oxide film up to a thickness of 1100 Å is shown in Figure 3-12 as the thin solid line. This takes us to the Del/Psi point of 80.72°/47.81°. This trajectory and the reference trajectory mentioned above are calculated using the one-film program listed in Appendix A. Then, using a two-film program, we calculate the trajectory for a growing nitride film. Zero thickness of the nitride film starts at the 80.72°/47.81° point and the trajectory continues from there, as shown in Figure 3-12 as the solid triangles, plotted at 10 Å increments. Although the period point is different for this two-film situation than for a one-film situation, the thickness at which the film closes on itself will

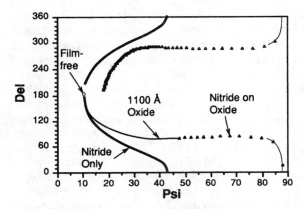

Figure 3-12. The trajectory for a two-film situation. Substrate is silicon. The next layer is 1100 Å of silicon dioxide, whose trajectory is shown as the thin solid line. On top of this is a growing silicon nitride layer shown by the solid triangles, plotted at 10 Å intervals. Also plotted as the thick solid line is the single-film trajectory for silicon nitride on silicon.

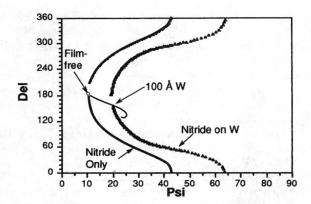

Figure 3-13. A trajectory for another two-film situation. The substrate is silicon. The next layer is 100 Å of tungsten, whose trajectory is shown as the thin solid line. On top of this is a growing silicon nitride layer shown by the solid triangles, plotted at 10 Å intervals. Also plotted as the thick solid line is the single-film trajectory for silicon nitride on silicon.

be the same value, 1972 Å, since the period thickness depends on the film only.

As another illustration, let us consider a silicon nitride film being deposited onto 100 Å of tungsten. In Figure 3-13, we show the trajectory for depositing tungsten on silicon as the thin solid line. In this case, we show the trajectory for 1300 Å. If we use the two-film program (three-film program with the thickness of the middle film set to zero) to calculate the trajectory for the deposition of silicon nitride on 100 Å of tungsten, the trajectory starts at the Del/Psi point of 156.3°/19.98° and proceeds from there. Again, the trajectory for the thickening nitride layer is shown as solid triangles plotted at 10 Å increments. The curve closes on itself at the point 156.3°/19.98°, again at a thickness of 1992 Å.

For multiple films, then, one starts the composite trajectory at the substrate point and traces out the first layer to its terminal thickness. Then, using the two-film program, one starts the second layer at this Del/Psi point and traces out the second layer to its terminal thickness. Successive programs can be used to continue this process *ad infinitum*. We list as appendices only the one-film and three-film programs. McCrackin's original program[3] allows up to 900 films. It should be noted that for the multiple-film programs, the thickness of the underlying layers is a required input. Errors in the underlying layer thickness

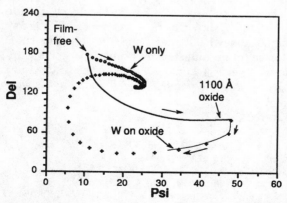

Figure 3-14. An example of a two-layer trajectory for an absorbing film on top of a dielectric film. In this case, tungsten on top of 1100 Å of silicon dioxide is shown in 10 Å increments.

values will cause the calculated trajectory to deviate significantly from its proper place. The use of targeted values for these thicknesses rather than actual measured values limits the number of layers that can be handled reasonably.

3.5.2 Absorbing Films on Top

As an example of a two-layer structure with an absorbing film on top, we use tungsten on top of 1100 Å of silicon dioxide. In Figure 3-14, we show, for reference, the trajectory for tungsten on silicon (also shown in Figure 3-10). As was indicated in the section on single absorbing films, the trajectory starts at the appropriate point for bulk substrate and must end at the appropriate point for bulk film material, in this case, at Del/Psi point 130.0°/23.3°.

For the two-film situation, we start at the substrate point for silicon at 178.5°/10.5° and trace the trajectory for the silicon dioxide film out to point 80.72°/47.81°. To do a two-film calculation, we use the three-film program listed in Appendix A (with the thickness of the middle film equal to zero) and calculate the trajectory for the growing tungsten film. As before, it must end up at the Del/Psi point for bulk film material, or in this case, 130.0°/23.3°.

The zero thickness location for the thickening tungsten film is the 1100 Å point for silicon dioxide, or 80.72°/47.81°. When we start adding tungsten, there is an abrupt change in direction, primarily due the rather large difference in the indices of refraction of silicon dioxide and tungsten. The trajectory is shown at 10 Å increments with the symbol (+). The trajectory is somewhat longer than for the one-film situation, hence the points for very thin tungsten are farther apart. As with the single-film situation, however, when the thickness has reached four values of 191 Å (see Section 3.4) the points are again indistinguishable.

3.6 References

1. F. L. McCrackin, E. Passaglia, R. R. Stromberg, and H. L. Steinberg, *J. Res. NBS A. Physics and Chemistry*, **67A**, 363 (1963).
2. A. N. Saxena, *J. Opt. Soc. Am.*, **55**, 1061 (1965).
3. F. L. McCrackin, *Nat. Bur. Stand. Tech. Note 479* (1969).
4. S. Chandrasekhar, A. S. Vengurlekar, S. K. Roy, and V. T. Karulkar, *J. Appl. Phys.*, **63**, 2072 (1988).

Chapter 4
Determining Optical Parameters for Inaccessible Substrates and Unknown Films

4. 1 Inaccessible Substrates and Unknown Films

In Chapter 3, we discussed the Del/Psi trajectories for various kinds of films. From that discussion, it might appear that one simply measures Del and Psi and then determines by inspection the value of the thickness. The Del/Psi trajectory can readily be calculated if the film-free values of Del and Psi are known (and hence $\tilde{N}_3 = n_3 - j\,k_3$) and if the index of refraction $\tilde{N}_2$ of the film is known. In fact, if these are known, one can usually use the canned software that came with the ellipsometer to determine thickness and often the index of the film.

We have discussed the danger of using handbook values for the index of refraction of the substrate and urged the reader to determine the values for the particular substrate of interest, although we have not discussed how one obtains these values. For a system such as an organic film deposited on gold, it is a simple matter to measure the surface before deposition, to obtain the values of Del° and Psi°, and then to measure the film-covered surface after deposition to obtain the values of Del and Psi. Some materials such as single crystal silicon can be fabricated in such a reproducible manner that the optical constants of the substrate can be stored in the microcomputer. In this case, the appearance is given that the thickness and index of refraction for the film is obtained from a single measurement.

As suggested in Figure 4-1, however, for many materials the optical properties of a film-free surface are not readily available. Metals,

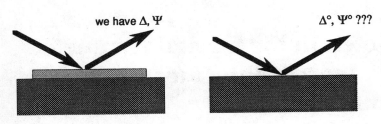

Figure 4-1. Del and Psi can always be measured for a film-covered surface. The film-free values of Del and Psi are not always accessible.

particularly, oxidize so readily that the oxide-free surface only exists in a vacuum. Ellipsometry can be done inside a vacuum system, but it is often not practical to construct this type of setup in order to obtain the necessary information. In this chapter, we will discuss how one can obtain useful information when the film-free substrate is inaccessible for measurement. Several examples from the author's work will be presented.

In this discussion, we use primarily ellipsometry, but we also use some other techniques for ancillary measurements. Specifically, Auger electron spectroscopy (AES),[1] x-ray photoelectron spectroscopy (XPS),[2] and Rutherford backscattering spectroscopy (RBS)[3] will be used in a supporting manner.

As mentioned above, we often do not know Del° and Psi°, the film-free values of Del and Psi. These values are affected by grain size, roughness, crystallinity, etc. We must either measure Del° and Psi°, as we might for an organic film on gold, we must know it from some other source, as with single crystal silicon, or we must <u>conjecture</u> what the values are. In some cases, one can etch the film off afterward and measure the resulting values of Del and Psi. (This author has found that such a method is not very reliable). In addition to not knowing the index of refraction of the surface (i.e., knowing Del° and Psi°), we may not know the values of the index of refraction of the film. The film index of refraction is, in general, a complex number. In some cases, the imaginary part is zero and the film is a pure dielectric.

If we have a single sample of film, conjecturing the film-free values of Del and Psi is rather risky. On the other hand, if one has a number of film samples that are believed to be the same composition and differing only in thickness, it is reasonable that we might be able to conjecture the film-free values. Having some additional information from ancillary techniques such as AES, XPS, and RBS gives a greater likelihood of success.

With respect to using ancillary techniques, a single ellipsometric measurement takes much less time than a single measurement from most ancillary techniques. If one only had one or two samples, one would probably use the existing ancillary method and not bother with developing the ellipsometry parameters. If one has many samples, however, the development of this method can save a sizable amount of time.

4.2 Determining Film-Free Values of Del and Psi

4.2.1 An Example Using One Method

We start with the example[4] of silicon dioxide on tungsten silicide. Tungsten silicide, or WSi_x, is deposited by chemical vapor deposition (CVD) using WF_6 and SiH_4. The stoichiometry is such that the value of x typically falls around 2.6. This structure has a higher resistivity than is required for normal integrated circuit interconnects and normally this material is annealed in nitrogen or oxygen to reduce the resistivity.[5-10] The annealing also changes the stoichiometry to reduce the ratio of silicon to tungsten, or the value of x, to a value nearer to 2.2. When the annealing is done in oxygen, some of the excess silicon reacts to form silicon dioxide on the surface.[11-13]

We often want to know how much oxide has formed, and often a rough estimate is all that is needed. Auger electron spectroscopy can always be used, but the measurements are rather time-consuming, and cannot be done in a wafer fabrication area. The following efforts were aimed at developing a way to use ellipsometry to obtain a thickness value that is accurate to within about 50 Å.

To develop the method, samples that had been submitted over a few months time for Auger analysis were also measured with an ellipsometer. The values of Del and Psi obtained are shown in Figure 4-2.

To obtain the thickness with ellipsometry, we must have the film-free values of Del and Psi. It is not simply a matter of measuring Del and Psi before the anneal since the anneal changes the grain structure and stoichiometry. Etching the oxide off and then measuring the values of

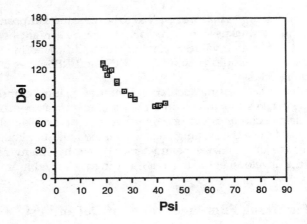

Figure 4-2. Del and Psi values for several examples of SiO_2 on WSi_x, where x ~ 2.2.

Del and Psi has been used by others,[7] but in the present case, the results were erratic and not dependable. We therefore will conjecture these values, using the information obtained from Auger electron spectroscopy. After the values are determined, subsequent samples will not have to be analyzed with AES.

Although, in general, Del can range from 0° to 360°, we showed in Section 3.1 that the values of Del° for a surface always fall between 0° and 180°. Similarly the values of Psi° always fall between 0° and 45°. Archer[14] shows and we observe from exercising the programs listed in Appendix A that the Del/Psi trajectory normally moves in a counter-clockwise direction on a graph such as Figure 4-2. From these two facts, we can conjecture that the film-free values of Del and Psi will fall in the region suggested in Figure 4-3.

Let us now use a trial-and-error approach and use Program OneFilm in Appendix A to calculate the Del/Psi trajectory for several possible values of Del° and Psi°. The first requirement is that the calculated trajectory must fall reasonably close to the observed data points. It turns out that there is a range of values of Del° and Psi° that will cause the calculated line to pass through the points. One such example is shown in Figure 4-4.

The second requirement is that the thickness measurements made with Auger electron spectroscopy along with argon ion etching must

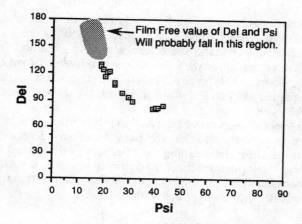

Figure 4-3. Postulated region where Del° and Psi° will be for the oxides of WSi$_x$.

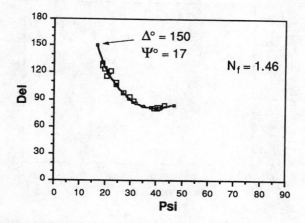

Figure 4-4. Calculated curve for the postulated values of Del° and Psi° shown along with the experimental data points.

also agree reasonably well with the thickness as determined from the calculated curves. The method then consists of choosing the postulated Del°/Psi° values that make the thickness values obtained from the calculated curve best agree with the thickness values obtained from the AES measurements. The "eyeball" best fit with these data was obtained with the postulated Del°/Psi° used in Figure 4-4 above and the fit is shown in Figure 4-5.

The expected accuracy of the AES depth profiling is within about 10%. Since we already had the AES data, we clearly do not need the ellipsometry data for determining the thickness of these samples. The value in this development is that for additional samples, the AES analysis is not necessary and the thickness of the oxide can be determined to within about 50 Å with ellipsometry alone.

4.2.2 An Example Using Another Method

Titanium nitride is a conductive material used in the microelectronics industries as a barrier between aluminum metallization lines and the active silicon.[15-17] Ellipsometry was used[18] to measure the oxidation of this material.

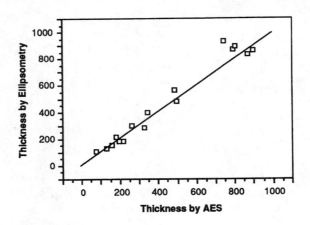

Figure 4-5. Thickness (in Å) determined by ellipsometry using Del° = 150° and Psi° = 17° compared with the thickness determined by AES with sputter profiling. The solid line represents exact fit.

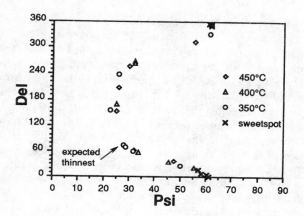

Figure 4-6. The Del/Psi values for TiN samples oxidized in air at the temperatures noted for times varying from 15 min to 42 hr.

Using XPS and RBS, we observed that the film which results from oxidation of TiN is TiO_2. The samples used in the kinetic study were oxidized at various times at the three chosen temperatures and the Del/Psi values were then measured. The values for the room air oxidation are shown in Figure 4-6. To determine accurately the value of the index of refraction of the film, it is important to have Del/Psi values near Del = 0° (or 360°). Accordingly, some samples that were excluded from the kinetic study because of uncertainty in the oxidizing temperature were included because the Del/Psi points advantageously fell near Del = 0. These are included in Figure 4-6 and are listed as "sweetspot." The data point labeled "expected thinnest" was the point for 350°C/1 hr.

To determine the thicknesses of the films using ellipsometry, we must generate the Del/Psi trajectory. To do this, we need to know the Del/Psi values for the TiN substrate (and hence the optical constants of the annealed TiN). It is not sufficient simply to measure the Del/Psi values for a film that has not been oxidized. The oxidation process is also an annealing process and will change the optical properties of the underlying TiN. In a previous work,[19] we observed a measurable difference in the optical constants of TiN films formed with substrate temperatures of 25°C and 325°C.

The film-free point on the Del/Psi plot will lie above and to the left of the expected thinnest point. To locate this point, we used an iterative method. For the first iteration, we assumed a point within 2° of both Del and Psi for the expected thinnest point. We then chose an approximate value for the index of refraction of the film (using methods discussed in the following section) and calculated a Del/Psi trajectory which gave a reasonable fit for most of the points. We then determined by inspection the thickness for the 350°C/42 hr point and the 450°C/2 hr point. The resulting thickness values were both slightly above 1000 Å. AES analysis along with argon ion depth profiling was done on these two films to determine the first iteration value of sputter rate in the AES system. The thickness of the thinnest film was then determined by AES depth profiling using the first iteration sputter etch rate. This Auger thickness value for the thinnest film was then used for the second iteration choice of the starting Del/Psi point and the process was repeated. Two iterations were sufficient to obtain values with accuracy commensurate with the rest of the study. It is estimated that the uncertainty in the zero-thickness point should cause no more than a 30 Å error in the thickness value obtained.

The following section discusses the determination of the index of refraction of the film. In addition, a third method of obtaining the film-free Del/Psi point will also be included.

4.3 Determining the Complex Index of Refraction of the Film

4.3.1 Description of the Materials for the Example

The example[20] we will use to illustrate the method of determining the complex index of refraction is the oxidation of the pseudoalloy Ti:W. This material is deposited by sputter deposition and typically has an atomic concentration of about 30% Ti and 70% W. It, too, is often used as a barrier between aluminum and silicon.[21-25] This material forms a native oxide which turns out to be about 20 Å thick. We heated several samples in oxygen for various times at temperatures of 375°C, 450°C, and 505°C for a total of 18 samples, which later turned out to have oxide thicknesses varying from about 100 Å to about 1000 Å. The Del/Psi points are shown in Figure 4-7.

Whereas in the case of WSi_x, the oxide is a familiar one, i.e., SiO_2, in this case, we know very little about the oxide that forms. AES analysis

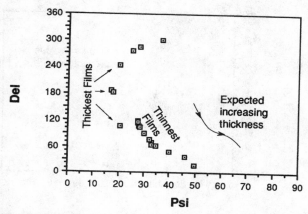

Figure 4-7. Experimentally observed Del/Psi values for Ti:W samples heated at various times at several temperatures. Values for the native oxides are also included.

showed that except for the first 20 Å or so, the film was compositionally uniform with depth. The AES analysis, x-ray photoelectron spectroscopy analysis, and Rutherford backscattering spectrometry analysis showed that the thicker oxide was probably a mixture of WO_3 and TiO_2. TEM analysis[26] suggested that the oxide was amorphous rather than crystalline.

4.3.2 Determining the Film-Free Point

Again, in the case of SiO_2, the sputtering rate can easily be calibrated, whereas the sputtering rate of the oxide of Ti:W is a different matter. In this case, we sputter etched one of the thicker samples with a knife-edge in place. We used AES to determine when the interface was reached. A stylus instrument was then used to measure the oxide thickness and to obtain a rough (± 30%) calibration of the sputtering rate. This rough calibration was then used with some of the very thin samples. A 30% error on a 1000 Å film will represent 300 Å whereas on a 20 Å native oxide, it only represents 6 Å. In the analysis described below, we only used the AES-determined thickness on the very thin films.

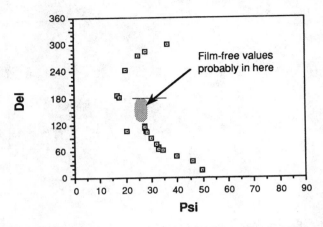

Figure 4-8. Postulated region where Psi° and Del° will be for the oxides of Ti:W.

Having a rough approximation of the thickness of the thinnest films, we simply backed off the Del°/Psi° point so that the calculation gave this value for these thinnest films.

4.3.3 The Film Index

Figure 4-7 shows the experimental data points obtained from the heated samples and the native oxides along with an indication of the expected increasing thickness (based on oxidation time and temperature). One of the first observations is that the Del/Psi trajectory does not appear to close on itself. The data points grouped at about Del = 120° and Psi = 27° are for very thin films whereas the data point at about Del = 105° and Psi = 20° is for a very thick film.

As in the case of the WSi_x, we can conjecture roughly where the Del°/Psi° point should be. It will probably lie between the data points for the thinnest films (the native oxides) and a value of 180°, probably in the region suggested by Figure 4-8.

One of the difficulties we now encounter is that in addition to not knowing the location of the starting point, we also do not know the index of refraction of the film. If we had only one or two data points, it would not be possible to determine this index. The fact that we have many points makes it reasonable that we can determine the index from the data themselves.

The fact that the Del /Psi curve does not close on itself indicates that there is a nonzero imaginary component of the film index of refraction, i.e., it is not pure dielectric. At this point, we choose our first iteration of the value of Del° and Psi° slightly above the value for the thinnest films. Using the modified McCrackin program, we calculate Del/Psi trajectories for several values of the real part of the index of refraction of the film. Using the best fit to the data for the part of the curve near Del = 0°, we add in a bit of the imaginary part of the index of refraction and recalculate the next iteration of the Del/Psi trajectory to determine the offset to the curve. Using this iterative method (trial-and-error) we obtain the best fit for the curve.

It turns out that the determination of the best fit for the value of the index of refraction is not particularly sensitive to the choice of Del° and Psi°. After making the best fit determination for the index of refraction, we reassess our choice of Del° and Psi°. We know the thickness of the thinnest films to within about 6 Å. The final choice of Del° and Psi° should be such that there is reasonable agreement between ellipsometry and AES for the thickness values of these thinnest films.

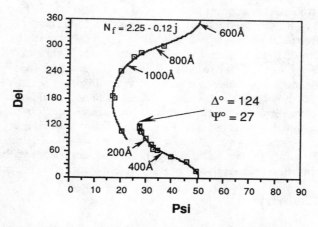

Figure 4-9. The Del/Psi trajectory that gives the best fit to experimental data and best thickness values for thin films.

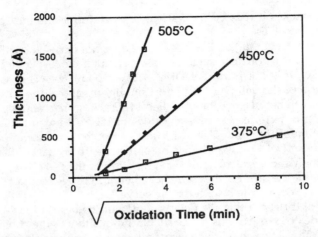

Figure 4-10. Thickness from ellipsometry vs. the square root of the oxidation time.

The best fit for the entire iterative process for these data is given in Figure 4-9 along with the appropriate index of refraction for the film.

The index of refraction of the substrate which corresponds to the given Del° and Psi° is $\tilde{N}_s = 2.84 - 3.08j$. The real part of the index of refraction of the film falls roughly in the area of TiO_2 and WO_3. The imaginary part of the index of the film suggests that this film is three to six times more absorbing than single crystal silicon.

To show that the thickness values obtained in this manner are consistent with what one might expect, we use the curve obtained above to determine the thicknesses of the various oxides and in Figure 4-10 we plot the thickness as a function of the square root of oxidation time for the three temperatures. The straight line fit is consistent with a diffusion-controlled process that is present in many forms of oxidation.

4.4   Summary

We have considered several cases where the optical parameters of the film-free sample were inaccessible. We have shown that with some information from an ancillary technique (e.g., AES) one can reasonably postulate the film-free values. We have suggested three possible ways of doing this.

The first way used samples with a range of thicknesses along with another accepted method of determining the thickness of these same samples. The best values for Del°/Psi° are those that give the best agreement between ellipsometry and the other technique. In the second and third method, a thicker sample is used to obtain a rough estimate of the sputter etch rate, which is in turn used to estimate the thickness of the very thin films. The error in the thickness of the thin films will be small, and the best Del°/Psi° point is that which makes the calculation give the best match for the thinnest films.

The complex index of refraction of the film was chosen in an iterative way to cause the calculated curve to fall through most of the points. For particularly accurate values, it is essential to have points with Del near 0°/360°.

4.5 References

1. "Auger Microprobe Analysis," I. F. Ferguson, Adam Hilger Publishing Co., New York (1989).
2. "Photoelectron and Auger Spectroscopy," T. A. Carlson, Plenum Press, New York (1975).
3. "Backscattering Spectrometry," W.-K. Chu, J. W. Mayer, M.-A. Nicolet, Academic Press, New York (1978).
4. H. G. Tompkins, *Thin Solid Films*, **181**, 285 (1989).
5. M. Kottke, F. Pintchovski, T. R. White, and P. J. Tobin, *J. Appl. Phys.*, **60**, 2835 (1986).
6. D. K. Sadana, A, E. Morgan, M. H. Norcott, and S. Naik, *J. Appl. Phys.*, **62**, 2830 (1987).
7. K. C. Saraswat, D. L. Brors, J. A. Fair, K. A. Monnig, and R. Beyers, *IEEE Trans. Electron Devices*, **ED-30**, 1497 (1983).
8. Y. Shioya and M. Maeda, *J. Appl. Phys.*, **60**, 327 (1986).
9. R. Togei, *J. Appl. Phys.*, **59**, 3582 (1986).
10. J. Kato, M. Asahina, H. Shimura, and Y. Yamamoto, *J. Electrochem. Soc.*, **133**, 794 (1986).
11. R. V. Joshi, Y. H. Kim, J. T. Wetzel and T. Lin, *Thin Solid Films*, **163**, 267 (1988).
12. D. K. Sadana, A. E. Morgan, N. H. Norcott, and S. Naik, *J. Appl. Phys.*, **62**, 2830 (1987).
13. L. Krusin-Elbaum and R. V. Joshi, *IBM J. Res. Develop.*, **31**, 634 (1987).
14. R. J. Archer. *J. Opt. Soc. Am.*, **52**, 970 (1962).
15. M. Wittmer and H. Melchior, *Thin Solid Films*, **93**, 397 (1982).

16. M. Kumar, J. T. McGinn, K. Pourrezaei, B. Lee, and E. C. Douglas, *J. Vac. Soc. Technol. A*, **3**, 2233 (1985).
17. N. Yokoyama, K. Hinode, and Y. Homma, *J. Electrochem. Soc.*, **136**, 882 (1989).
18. H. G. Tompkins, *J. Appl. Phys.*, **70**, 3876 (1991).
19. H. G. Tompkins, R. Gregory, and B. Boeck, *Surface and Interface Analysis*, **17**, 22 (1991).
20. H. G. Tompkins and S. Lytle, *J. Appl. Phys.*, **64**, 3269 (1988).
21. P. B. Ghate, J. C. Blair, C. R. Fuller, and G. E. McGuire, *Thin Solid Films*, **53**, 117 (1978).
22. P. Merchant and J. Amano, *J. Vac. Sci. Technol. A*, **1**, 459 (1983).
23. C. Y. Ting and M. Witmer, *Thin Solid Films*, **96**, 327 (1982).
24. R. S. Nowicki and B. Schiefelbein, in Proceedings, Tungsten & Other Refractory Metals for VLSI Applications, edited by R. S. Blewer (Materials Research Society, Pittsburgh, PA (1986).
25. T. Hara, N. Ohtsuka, K. Sakiyama, and S. Saito, *IEEE Trans. Electron Devices*, **ED-34**, 593 (1987).
26. P. L. Fejes, private communication (1988).

Chapter 5
Extremely Thin Films

5.1 General Principles

Let us rather arbitrarily define an "extremely thin" film as a film less than 50 Å thick. In this category fall such films as native oxides, adsorbed monolayers, and Langmuir-Blodgett films. From Figure 5-1, we see that the trajectories for the various film indices of refraction are quite close together. From this we can surmise that we will not be able to use the Del/Psi measurements to obtain information about the index of refraction of the film. Index of refraction is still important, however,

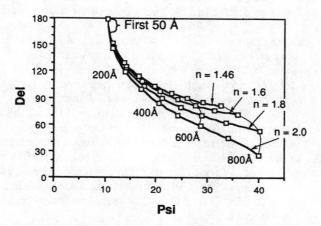

Figure 5-1. The Del/Psi trajectory for film with various indices on silicon. Note that the trajectories for the first 50 Å are not particularly well separated.

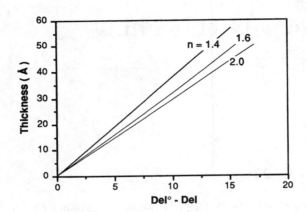

Figure 5-2. The variation of thickness with change in Del for extremely thin films on silicon for the indices of refraction shown.

since it affects the scale of our "yardstick." Figure 5-2 shows the relationship between changes in Del and the thickness as a function of the index of refraction.

For the substrate used in the example of Figure 5-1, the value of Psi does not vary significantly for extremely thin films. In some cases (e.g., for tungsten), both Del and Psi vary.

One might ask how thin a film can be measured with ellipsometry. Note that in the example shown in Figures 5-1 and 5-2, for an index of refraction of 2.0, a change in Del of 0.25° corresponds to a thickness change of 1 Å. Since we can measure values of Del to accuracies of less than 0.1°, it is reasonable to expect to be able to measure fractions of a monolayer. In practice, the limitation is the control of the surface rather than the measurement. When extremely thin films are to be measured, spot-to-spot variation in Del and Psi may well be more than the expected change in these quantities. Accordingly, most measurements of change in thickness must be made *in situ*.

Early ellipsometric measurements were made on very thin films. In 1890, Drude[1] observed films on polished metal surfaces only a few angstroms thick. The Drude approximation[2] for thin films is that the change in both Del and Psi is linearly related to the change in thickness. In equation form, this is

$$\Delta - \Delta^\circ = C_\Delta x \tag{1}$$

and

$$\Psi - \Psi^\circ = C_\Psi x \tag{2}$$

where x is the thickness and C_Δ and C_Ψ are constants.

The method of determination of the values of the constants depends on the film/substrate system being investigated. When there is some sort of internal calibration such as with Langmuir-Blodgett films or with the chemisorption of monolayers, the values of the constants can be determined empirically even though the index of refraction of the film

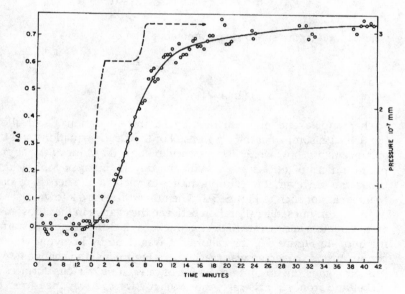

Figure 5-3. The change in Δ during chemisorption of oxygen on cleaved silicon under high vacuum at room temperature versus exposure time. After an initial steep rise, $\delta\Delta$ saturates when a complete monolayer has been formed. Also shown by the dashed curve is the rise in pressure as O_2 was suddenly admitted. (After Archer[3])

material may not be known. An example of this is shown in Figure 5-3 where Archer[3] shows that during chemisorption of oxygen on cleaved silicon, Del changes approximately 0.7° upon formation of a complete monolayer. Clearly, a change of 0.35° would represent a half-monolayer.

If the index of refraction of the film material is known, the constants C_Δ and C_Ψ can be evaluated by equations given by Drude[1,2]:

$$C_\Delta = \left(\frac{180}{\pi}\right)\left(\frac{4\pi}{\lambda}\right)\frac{\cos\phi_1\sin^2\phi_1(\cos^2\phi_1-\alpha)(1-1/n_2^2)}{(\cos^2\phi_1-\alpha)^2+\alpha_1^2} \qquad (3)$$

and

$$C_\Psi = \left(\frac{180}{\pi}\frac{\sin 2\Psi}{2}\right)\left(\frac{4\pi}{\lambda}\right)\frac{\cos\phi_1\sin^2\phi_1\cdot\alpha_1(1-n_2^2\cos^2\phi_1)(1-1/n_2^2)}{(\cos^2\phi_1-\alpha)^2+\alpha_1^2} \qquad (4)$$

where α and α_1 are substrate parameters given by

$$\alpha = (n_3^2 - k_3^2)/[(n_3^2 + k_3^2)^2],$$

and

$$\alpha_1 = (2n_3k_3/[(n_3^2 + k_3^2)^2].$$

Archer[4] and Saxena[2] give enhancements to these equations which allow the use with somewhat thicker films. For the Drude equations and the enhancements, however, it is necessary to know the value of the appropriate indices of refractions. With the development of McCrackin's program,[5] however, it is a simple matter to calculate the necessary data to make a plot such as Figure 5-2, when the indices are known.

Oftentimes, the calibration between the change in the ellipsometric parameters and coverage or film thickness is done using an external method. In Figure 5-3, the calibration was done by knowing when a certain surface coverage was reached. In Figure 5-4, the change in Δ is plotted versus the absolute coverage of oxygen on various surfaces of nickel and iron. In this case, the oxygen coverage was measured by nuclear reaction analysis (NRA). In some of these cases, the formation of thin oxide layers is occurring. Recall that a monolayer of atoms amounts to coverages of the order of 10^{15} atoms/cm^2.

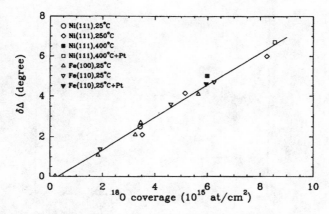

Figure 5-4. Ellipsometric parameter δΔ versus the absolute oxygen coverages on various surfaces of nickel and iron, both clean and covered with a small amount of platinum, as measured with the $^{18}O(p,\alpha)^{15}N$ nuclear reaction. The solid line is a least-squares fit to the data. (After Deckers[6])

5.2 Some Examples of Extremely Thin Films

5.2.1 Oxygen and Carbon Monoxide Adsorption on Silver

Albers *et al.*[7] report a study of the adsorption of oxygen on Ag(110) and the effect of subsequent exposure to CO. These studies are aimed at providing background information with regard to catalytic oxidation of ethylene. The work is done with ellipsometry, low-energy electron diffraction (LEED), and work function changes.

Experimental Methods

The experiments were done in an ultrahigh vacuum (UHV) system equipped with LEED and ellipsometry. The silver sample was spark-cut to within 1° of the (110) orientation from a 6N purity silver rod. It was then ground, electro-lap-polished, and mounted in the sample holder. Cleaning was done by sputtering followed by annealing in

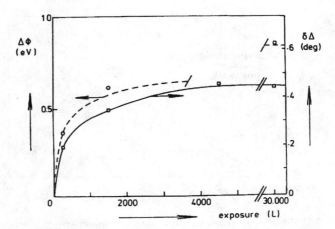

Figure 5-5. Changes in work function (circles) and Δ (squares) for various exposures of oxygen to Ag(110) at room temperature. The dotted line is taken from Ref. 9. (After Albers[7])

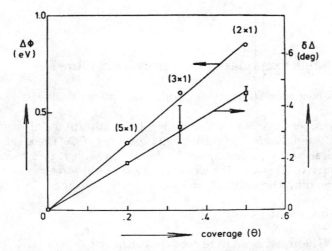

Figure 5-6. Work function changes (Δφ) and ellipsometric parameter changes (δΔ) as a function of oxygen coverage on Ag(110) as derived from LEED patterns. For the (3x1) structure the error bar indicates the range of δΔ for which this structure was visible; for the (2x1) structure the error in δΔ observed upon saturation. (After Albers[7])

UHV. To avoid unintentional interaction with CO, hot filaments were excluded as far as possible and titanium gettering was used regularly. Changes in the ellipsometric parameters ($\delta\Delta = \Delta° - \Delta$, $\delta\Psi = \Psi° - \Psi$) were measured with off-null irradiance measurements at an angle of incidence of 71° and a wavelength of 6328 Å. Work function changes were measured using the LEED apparatus.[8]

Results

Figure 5-5 shows the change in the work function ϕ and the ellipsometric parameter Δ upon exposure to oxygen at pressures up to 4×10^{-5} Torr with the crystal at room temperature. At exposures of about 6000 L, the (2x1) LEED structure is clearly visible. At higher exposures, $\delta\Delta$ becomes virtually constant. With some assumptions[7] about coverage and LEED patterns, the relationship between coverage, θ, and $\delta\Delta$, which is shown in Figure 5-6, was determined. The relationship can be expressed as

$$\theta = (0.92 \pm 0.05)\delta\Delta \qquad (5)$$

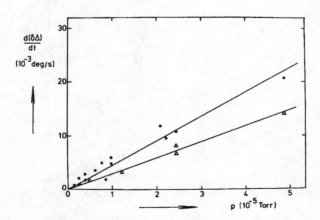

Figure 5-7. The initial rate of oxygen adsorption [d(δΔ)/dt] versus oxygen pressure: (filled circles) room temperature, (open triangles) at 469 K. (After Albers[7])

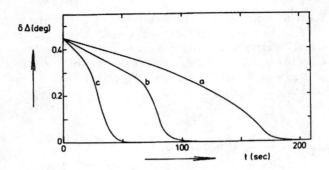

Figure 5-8. Change in $\delta\Delta$ as function of time after admission of CO at room temperature: (a) $p_{CO} = 3.7 \times 10^{-7}$ Torr; (b) 10^{-6} Torr; (c) 1.9×10^{-6} Torr. (After Albers[7])

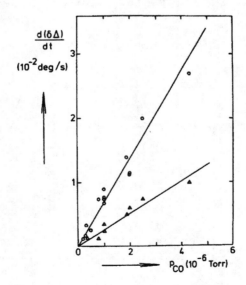

Figure 5-9. The initial (triangles) and "midway" slopes (circles) of curves as shown in Figure 5-8 as a function of CO pressure at room temperature. (After Albers[7])

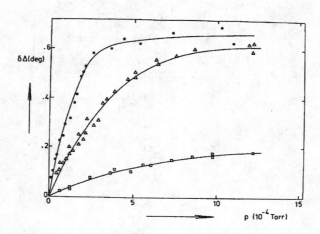

Figure 5-10. Adsorption isotherms of oxygen on Ag(110) at 542 K (dots), 550 K (triangles), and 583 K (squares) expressed as the change in Δ against oxygen pressure. (After Albers[7])

The rate of oxygen adsorption was studied at temperatures between 294 and 475 K in the 10^{-6} to 10^{-5} Torr range. Figure 5-7 shows rate measurements versus oxygen pressure. The conclusions were that the initial adsorption rates were proportional to the gas pressure and the adsorption curves could be described by an equation of the form

$$d\theta/dt = b(\theta_{max} - \theta) \qquad (6)$$

with θ_{max} independent of temperature.

The reaction of adsorbed oxygen with gaseous CO was studied in the same temperature range with CO pressures between 10^{-7} and 4×10^{-6} Torr. Rather complicated reaction kinetics were observed. Figure 5-8 shows the results. The effect of the CO is to remove the oxygen, and the interaction is most effective when the adsorbed oxygen atom is by itself, i.e., not surrounded by other adsorbed oxygen atoms. Because, initially, the oxygen coverage is high, the removal is slow. As

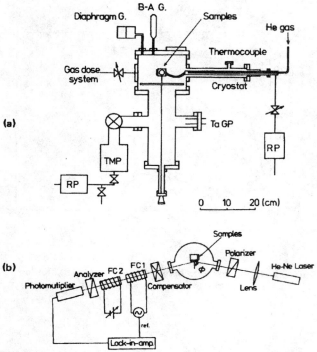

Figure 5-11. (a) Schematic diagram of the experimental apparatus. (b) Schematic diagram of the reflection ellipsometry system. Here FC1 and FC2 are Faraday cells. (After Itakura[10])

the removal proceeds, more lone oxygen atoms become available and the rate of removal increases. Eventually, the population decreases significantly. Figure 5-9 shows the initial and midway slopes of the curves shown in Figure 5-8 as a function of CO pressure.

The above observations dealt with how fast adsorption occurred, i.e., the kinetics. Figure 5-10 shows equilibrium coverages, or the adsorption isotherms of oxygen adsorbing onto Ag(110).

5.2.2   Adsorption of Xe on Ag(111)

Itakura and Arakawa[10] describe the measurement of the physical adsorption of Xe on Ag(111) measured with ellipsometry. The ellipsometer used was an *in situ* system rather than a turnkey system. A He-Ne laser at 6328 Å was used and the angle of incidence was 74°. Faraday cells[11] along with a lock-in amplifier were used to enhance the sensitivity. The schematic setup is shown in Figure 5-11.

The Ag samples were held in an aluminum vacuum system with base pressure below 10^{-7} Pa, pumped with a turbomolecular pump and a tantalum getter pump. Windows of the vacuum chamber were disks of fused quartz glass, mounted on two Viton O-ring seals. The Xe pressure, during exposure, was maintained by the dynamic balance between the gas flow rate and pumping speed of the cold surfaces of the cryostat at about 30 K. The sample was mounted on the cryostat and the desired temperature was controlled by electric heaters inserted in the sample holder.

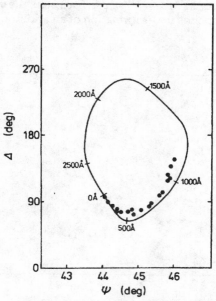

Figure 5-12. The Δ-Ψ diagram of Xe film on Ag(111) with the layer thickness as a parameter. The solid line represents the calculated result for optical constants 0.126 - 3.49 j for Ag(111) and 1.48 - 0.0 j for Xe. The angle of incidence is 74°. (After Itakura[10])

The crystal surfaces were cleaned by Xe ion bombardment followed by annealing at about 700 K. This procedure was repeated until no change was observed in the ellipsometric parameters.

Prior to the isotherm measurements, the Del/Psi trajectory was determined by adsorption of multilayers of Xe at pressures slightly higher than the saturation vapor pressure at 50 K. Multilayers up to about 1000 Å were grown with Del and Psi measured at various thicknesses. The optical constants of the substrate were determined by measuring the clean surface. The optical constants of the Xe film were determined in much the same manner as described in Chapter 4. Figure 5-12 shows a typical Δ/Ψ trajectory. The best fit curve resulted from optical constants of $\tilde{N}_s = 0.126 - 3.49\,j$ for the Ag(111) substrate $\tilde{N}_f = 1.48 - 0.0\,j$ for the Xe film. The trajectory also indicates that for a few monolayers, only Del varies significantly.

They made isotherm measurements at several different temperatures between 50 and 70 K. We show the results in Figure 5-13. The isotherms of Xe on Ag have several steps. Each step, the vertical part in the isotherm, corresponds to the formation of an additional monolayer.

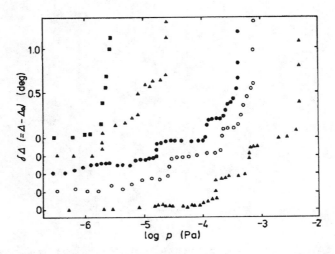

Figure 5-13. Ellipsometric isotherms of Xe on Ag(111) obtained by using the Faraday cells at the temperature of 50.0 K (filled squares), 57.8 K (open triangles), 63.5 K (filled circles), 64.0 K (open circles), and 65.2 K (filled triangles). (After Itakura[10])

The authors suggest that this is due to a first-order phase condensation where 2-D gas and 2-D solids coexist. They indicate that the gradual shifts of δΔ observed in the submonolayer region were due to the heterogeneity of the surface on the original Xe adsorbed. For the 63.5 K isotherm, layer-by-layer growth can be observed for three distinct layers. As Xe pressure approaches the saturation vapor pressure, δΔ increases rapidly because the film grows as 3-D bulk. The height of the first step is about 0.4°.

In addition, Itakura and Arakawa[10] also study the adsorption of Xe on Pt(111) and make various comparisons and contrasts to Xe on Ag(111). We have used only the Xe on Ag(111) results to illustrate the relevant ellipsometry aspects. This work observes the formation of up to three monolayers, one at a time. Although they did not use a turnkey system, the changes in Del were in the tenths of a degree range, which is well within the range of turnkey systems.

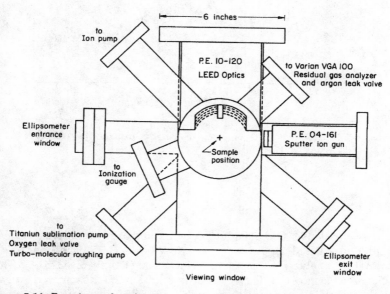

Figure 5-14. Experimental configuration. Ion pump, ionization gauge, and RGA are mounted 8.25 cm above the plane of the sample. (After Christensen[12])

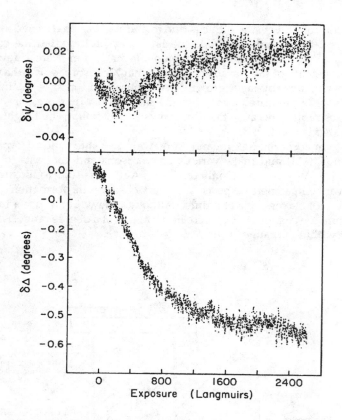

Figure 5-15. Ellipsometry data for a room temperature oxidation of Be(0001) at 1×10^{-7} Torr. (After Christensen[12])

5.2.3 Oxidation of Beryllium

Christensen and Blakely[12] report a study of the initial stages of the oxidation of Be(0001). This work was done under UHV conditions using a rotating analyzer ellipsometer with a wavelength of 6328 Å. The schematic is shown in Figure 5-14. In addition to ellipsometry, Auger analysis, LEED, and residual gas analysis could be done on the system or sample.

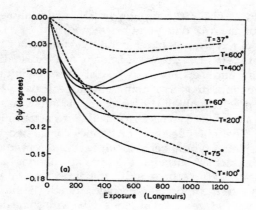

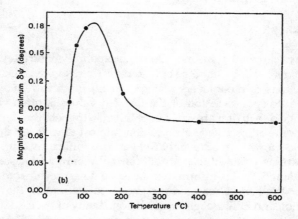

Figure 5-16. (a) Change in Ψ during Be(0001) oxidation at various temperatures. The magnitude of the decrease in Ψ passes through a maximum around 100°C. This is illustrated in (b) where the magnitude of the maximum δΨ at each oxidation temperature is plotted against the temperature. The decrease in Ψ is believed to be related to the incorporation of oxygen into the Be surface. (After Christensen[12])

A beryllium single crystal was spark-cut along the (0001) face and mechanically polished. Cleaning was done by sputtering and annealing in UHV. The cleanliness was monitored by AES. The ellipsometer was controlled by a DEC PDP 11/03 computer system, which allowed data acquisition at about one Δ/Ψ data point per second.

The results of a typical room temperature oxidation of Be(0001) are shown in Figure 5-15. Del decreases initially as the oxidation proceeds, rapidly at first and then more slowly. Psi decreases initially and then increases. The simple model[1] of a single homogeneous layer on top of a bulk material predicts only the increase, not the initial decrease.

Figure 5-16 shows the $\delta\Psi$-curves for oxidations at various different temperatures. The initial decrease in Ψ, once considered "anomalous," is now considered typical behavior since it is observed on most metal[13] and semiconductor[14] surfaces. Various explanations have been given for this behavior. The source of this initial decrease in Ψ on Be appears[12] to involve the incorporation of oxygen into the Be surface and the formation of a mixed layer of BeO islands in Be with some depletion of the Be valence electron density.

5.3 Summary

We have shown that the measurement capability of turnkey ellipsometer systems is adequate to be able to detect films with thicknesses down to fractions of monolayers. When dealing with extremely thin films, the spot-to-spot variation of the Del/Psi values is often greater than the change due to film growth, hence when extremely thin films are to be studied, the ellipsometry observations almost always are done *in situ*. We have shown that often only Del varies significantly during the growth of extremely thin films and that normally only thickness information is obtained. Index information must be determined from other experiments. We have illustrated the growth of extremely thin films with adsorption and initial stages of oxidation. Many of the case studies found later in the book illustrate other examples of very thin films. The interaction of benzotriazole with copper in solution is an interesting nonvacuum example.

5.4 References

1. P. Drude, *Ann. Phys. Chem.*, **36**, 865 (1889).
2. A. N. Saxena, *J. Opt. Soc. Am.*, **55**, 1061 (1965).
3. R. J. Archer, "Ellipsometry in the Measurement of Surfaces and Thin Films," edited by E. Passaglia, R. R. Stromberg, and J. Kruger, *Natl. Bur. Std, Misc. Publ. 256* , U. S. Government Printing Office, Washington (1964), pp. 255 ff.

4. R. J. Archer, *J. Electrochem. Soc.*, **104**, 619 (1957); R. J. Archer and G. W. Gobeli, *J. Phys. Chem. Solids*, **26**, 343 (1965).
5. F. L. McCrackin, *Nat. Bur. Stand. Tech. Note* **479** (1969).
6. S. Deckers, F. H. P. M. Habraken, W. F. van der Weg, and J. W. Geus, *Appl. Surf. Sci.*, **45**, 207 (1990).
7. H. Albers, W. J. J. Van der Wal, O. L. J. Gijzeman, and G. A. Bootsma, *Surface Sci.*, **77**, 1 (1978).
8. J. J. Lander and J. Morrison, *J. Appl. Phys.*, **34**, 3517 (1963); A. U. MacRea, *Surface Sci.*, **1**, 319 (1964).
9. H. A. Engelhardt and D. Menzel, *Surface Sci.*, **57**, 591 (1976); H. A. Engelhardt, A. M. Bradshaw and D. Menzel, *Surface Sci.*, **40**, 410 (1973).
10. A. Itakura and I. Arakawa, *J. Vac. Sci. Technol. A*, **9**, 1779 (1991).
11. H. J. Mathieu, D. E. McClure, and R. H. Muller, *Rev. Sci. Instrum.*, **45** 798 (1974).
12. T. M. Christensen and J. M. Blakely, *J. Vac. Sci. Technol. A*, **3**, 1607 (1985).
13. F. H. P. M. Habraken, O. L. J. Gijzeman, and G. A. Bootsma, *Surface Sci.*, **96**, 482 (1980); B. H. Heyden, E. Schweizer, R. Kotz, and A. M. Bradshaw, *Surface Sci.*, **111**, 26 (1981); J. Grimblot and J. M. Eldridge, *J. Electrochem. Soc.*, **129**, 2366 (1982); T. M. Christensen and J. M. Blakely, *J. Phys. Colloque*, **44**, C10 (1983).
14. F. Meyer, *Surface Sci.*, **56**, 37 (1976).

Chapter 6
The Special Case of Polysilicon

6.1 General

In silicon-based integrated circuits, the transistors, diodes, and other active elements are fabricated in either the original single crystal silicon wafer or in epitaxially grown silicon on top of the original wafer. These are connected to each other and to the bond pads (for connection to the outside world) with interconnecting lines, often made of aluminum. Sometimes, however, it is preferable to connect the devices with doped polycrystalline silicon[1] instead. This material is also used as a gate electrode, dielectric isolation, etc. This material is deposited by a chemical vapor deposition process where silane, SiH_4 is heated and decomposes to form a silicon layer. The crystalline structure of the layer depends on the deposition parameters[2] and can vary from amorphous to polycrystalline. This material is called "polysilicon," which is short for "polycrystalline silicon," although the descriptive term "amorphous polysilicon" is clearly an oxymoron. In general, the term "polysilicon" reveals more about the process used for the deposition than it does about the crystalline nature of the material.

6.2 Range of the Optical Constants

The generally accepted values for the index of refraction for single crystal silicon is about 3.87 ± 0.02. The value of the extinction coefficient is usually quoted as 0.025 ± 0.01. The extinction coefficient determined from ellipsometric measurements is often subject to error due to roughness, the presence of a thin contamination film, etc. The value reported by NIST[3] is $k = 0.015464 \pm 0.00031$. For the most part, these differences are insignificant and, in many cases, the optical constants of single crystal silicon are preprogrammed into the software of turnkey ellipsometers.

One would expect the optical constants of polysilicon to be near those of single crystal silicon. Although the values for polysilicon are not too far removed from those of single crystal silicon, the variations due to deposition or annealing cause some profound difficulties. Hirose *et al.*[4] did some early work on measuring the values of n and k at various wavelengths. They deposited the samples on quartz with the films varying in thickness from 0.5 μm to 1.5 μm. The optical properties were measured by photometry, using transmission. They observed that the samples were crystalline when deposited at temperatures above 675°C, and that when samples deposited at lower temperatures were annealed above 690°C they transformed from amorphous to polycrystalline. All of the films used for the n and k measurements were deposited as amorphous. Some were then annealed 1050°C to form the polycrystal-line films. The values obtained for a wavelength of 6328 Å are listed in Table 6-1. Clearly, the value for k for polycrystalline silicon is similar to that for single crystalline silicon, whereas that for amorphous silicon is significantly greater.

Jones and Wesolowski[5] deposited polycrystalline films that were 0.45 μm thick and were doped to strongly degenerate concentrations ($\sim 2 \times 10^{20}$ cm^{-3}) by thermal diffusion of phosphorus. They measured the optical constants by transmission photometry at various wavelengths

TABLE 6-1. Values of Absorption Coefficient, Extinction Coefficient, and Index of Refraction for Various forms of silicon. (from Hirose *et al.*[4])

	α	k	n
single crystal Si	3.9×10^3 cm^{-1}	0.019	
polycrystalline Si	6.9×10^3	0.035	~4.0
amorphous Si	3.8×10^4	0.192	~4.3

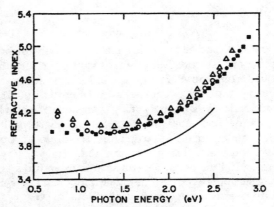

Figure 6-1. Index of refraction spectral curves for polysilicon films. Symbols denote the different processing parameters as follows: (open circles) undoped and unannealed; (triangles) undoped but annealed; (solid circles) 0.94×10^{20} cm^{-3}; (squares) 1.72×10^{20} cm^{-3}. Single crystal data are from H. R. Phillipp and E. A Taft, *Phys. Rev.*, **120**, 37 (1960) and C. Salzberg and J. Villa, *J. Opt. Soc. Am.*, **47**, 244 (1957). (After Jones[5])

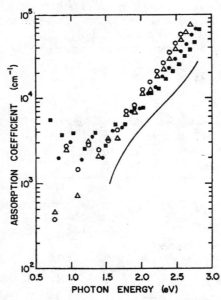

Figure 6-2. Absorption coefficient α as a function of wavelength for polysilicon films. Symbols are as in Figure 6-1. Single crystal data are from W. C. Dash and R. Newman, *Phys. Rev.*, **99**, 1151 (1955). (After Jones[5])

covering some of the UV and all of the visible wavelengths. Figure 6-1 shows their experimental observations for n as a function of photon energy. The wavelength of 6328 Å corresponds to a photon energy of 1.96 eV. They make the observation that "Within the experimental error it is difficult to observe any shift in the index of refraction with annealing or doping." Figure 6-2 shows their experimental observations for the absorption coefficient α as a function of photon energy. They also observe that "At high photon energies we see that there clearly is a dependence of α on processing. With annealing or increasing carrier concentration, there is a blue shift in the absorption curve. However, since both high temperature annealing and heavy P doping are known to enhance grain growth, this shift may be due to grain growth."

A later study was that of Chandrasekhar.[2] In this work polycrystalline Si films are deposited onto optically flat quartz wafers in a commercial LPCVD system at various substrate temperatures. The disorder in the films was characterized with x-ray diffraction and Raman scattering. The optical constants were measured with ellipsometry at two angles of incidence. They ignore the effect of any roughness and also the thin native oxide that grows during the transit time between deposition and ellipsometry measurements.

They find that films deposited at 580°C are x-ray and Raman amorphous. Films deposited at 600 and 620°C have both a microcrystalline component and an amorphous component with the 600°C sample being mostly amorphous and the 620°C film being mostly polycrystalline. Samples deposited at 640°C have an average grain size of about 400 Å and no significant amorphous component. They report that samples deposited at lower temperatures and then heated to 700°C showed no evidence of any structural modification.

The optical constants they reported are plotted in Figure 6-3. The values of n and k for the 640°C samples are not that far from those of single crystal silicon. When a significant amorphous component is present, however, the values of both n and k increase significantly.

In general, workers in the integrated circuit realm have an advantage that workers in many other discipline (corrosion science, metallurgy, etc.) do not have. Materials such as single crystalline silicon, silicon dioxide, silicon nitride, etc., can be produced in a very reproducible manner. Accordingly, the optical constants for these materials do not vary significantly from lot to lot. Unfortunately, the material polysilicon is more typical of other disciplines rather than the integrated circuit realm. The first difficulty then is that the optical constants of

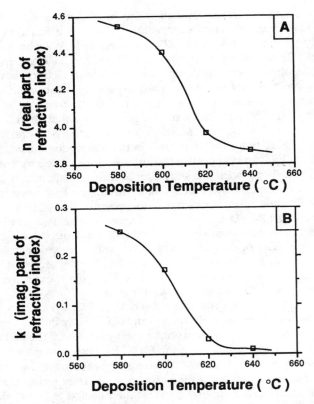

Figure 6-3. Values of the optical constants for polycrystalline silicon versus the sample temperature during deposition. Data from Chandrasekhar.[2] (A) The value of n. (B) The value of k.

polycrystalline Si depend on the processing conditions, and the variations are significant even within the range of normal process conditions.

6.3   Del/Psi Trajectories in General

Polysilicon is almost invariably deposited on silicon dioxide. The starting point for the polysilicon Del/Psi trajectory then depends on the thickness of the oxide. Using optical parameters similar to those used

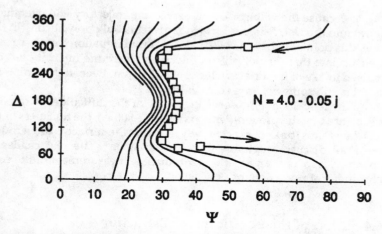

Figure 6-4. The calculated Del/Psi trajectory for polysilicon on 1000 Å of silicon dioxide on silicon. The index of refraction for the oxide is Ñ = 1.46 and for the polysilicon is Ñ = 4.0 - 0.05 j.

by Irene and Dong,[6] we plot the trajectory for Figure 6-4 showing a trajectory for polysilicon on 1000 Å of silicon dioxide on single crystal silicon. The index of refraction for the silicon dioxide is taken to be Ñ = 1.46 - 0.0 j. The single crystal silicon starts at a Del/Psi point of about 178.5°/10.5°, and the addition of 1000 Å silicon dioxide takes the Del/Psi point to about 79.5°/41°. This part of the trajectory is not shown in Figure 6-4. We start the polysilicon trajectory at this point. For this example, the index of refraction for the polysilicon is taken to be Ñ = 4.0 - 0.05 j. We observe that the regions of the trajectory near Del = 0°/360° are reasonably well separated, but the regions of the trajectory near Del = 180° are quite close together. Whereas the curve for a dielectric (k = 0) closes on itself after one period (see Section 3.3), the curve for polysilicon is offset slightly. The reason that it nearly closes is due to the fact that for polysilicon, k is very small. If we consider the "pseudo-period" to be the thickness interval between successive times when the trajectory passes Del = 180°, the pseudo-period is 814 Å. This can be calculated from Equation 4 of Chapter 3 using the real part of the refractive index.

Because the value of k is not zero, the trajectory will eventually spiral into the Del/Psi point, which represents bulk polysilicon. In this case, this point will be at 178.20°/11.43°. From Equation 8 of Chapter 1, we calculate that the polysilicon thickness where the intensity has decreased to $1/e$ of its surface value is about 10 μm. Even at this point, the spiral is still some distance from its center.

Irene and Dong[6] observe that some of the difficulties with polysilicon are that the pseudo-period is small (~ 800 Å), the thickness of the SiO$_2$ layer must be known accurately, and that the surface is often rough. We shall discuss roughness in a later chapter. The real challenge, however, is to have an accurate reproducible measurement of the complex index of refraction for polysilicon.

6.4   Effect of the Coefficient of Extinction

The effect of changing the coefficient of extinction k is to change the location of the Del/Psi trajectory. Figure 6-5 illustrates this point. In this case, we plot the trajectory of polysilicon on 530 Å of silicon dioxide for three different values of the extinction coefficient for polysilicon. Note that in these plots, the entire range of Del is shown, but we have used only one-third of the range of Psi, hence the shapes of the curves are somewhat skewed.

It would be reasonably straightforward to determine thicknesses and the index of refraction if one had samples with several different thicknesses. This is particularly true if some of the Del/Psi points fell in the "sweetspot" near Del = 0°/360°. Often, however, one does not have a range of thickness values. Consider the dilemma of having a Del/Psi value of 180°/28.15° when the index of refraction is not known particularly accurately. From Figure 6-5A, one would conclude that the sample is 2140 Å thick. From Figure 6-5B, the value would be indeterminate since the point falls between the curves and from Figure 6-5C, one would conclude that the film was only 510 Å thick.

6.5   Effect of the Index of Refraction

Whereas the effect of changing the coefficient of extinction k is to change the location of the trajectory, the effect of changing the index of

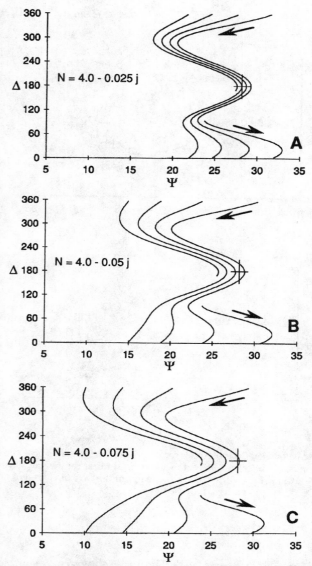

Figure 6-5. Calculated Del/Psi trajectories for polysilicon on 530 Å of silicon dioxide, illustrating the effect of different values of k. Refractive indices for the polysilicon are (A) $\tilde{N} = 4.0 - 0.025\,j$, (B) $\tilde{N} = 4.0 - 0.050\,j$, and (C) $\tilde{N} = 4.0 - 0.075\,j$. The arrows indicate the trajectory direction with increasing thickness of polysilicon. The cross-hairs represent the Del/Psi point 180°/28.15° in all three parts of the figure.

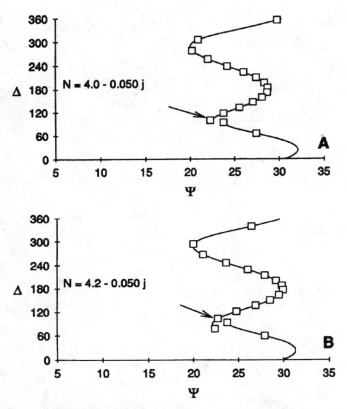

Figure 6-6. Calculated Del/Psi trajectories for polysilicon on 530 Å of silicon dioxide, illustrating the effect of different values of n. Refractive indices for the polysilicon are (A) $\tilde{N}$ = 4.0 - 0.050 j, (B) $\tilde{N}$ = 4.2 - 0.050 j.　　The tips of the arrows represent the Del/Psi point 99.5°/22.3° in both parts of the figure.

refraction n is to change the yardstick. Figure 6-6 shows two examples of trajectories using different values of n. In this figure, the squares are plotted at 50 Å intervals. A Del/Psi value at the location indicated by the arrow would give a thickness value of 800 Å if the value of n were 4.0, but would give a value of 760 Å if n were 4.2.

6.6 Requirements

If single-angle single-wavelength ellipsometry is to be used effectively to measure the thickness of polysilicon, the following requirements must be met:

- The thickness of the underlying oxide must be known reasonably accurately. An error of 30 Å on a 530 Å oxide film will cause the measured Del/Psi points to fall between one pseudo-period and the next pseudo-period of the calculated trajectory. An error of 50 Å will cause the point to fall on the wrong pseudo-period.
- The index of refraction of the polysilicon must be known.
- For self-consistency, a rough estimate (within 800 Å) of the thickness of the polysilicon should be known in order to assure that the proper pseudo-period was used.

6.7 Measuring the Thickness of Oxide on Polysilicon

At first glance, it might seem that since we so readily use ellipsometry to measure the thickness of silicon dioxide on silicon, we should be able to use ellipsometry to measure the thickness of silicon dioxide on polysilicon. In actual practice, this is very complicated. The extinction coefficient for polysilicon is larger than that of single crystal silicon, but compared to metals, it is still quite small. This has the effect that the light interacts with the layers below the polysilicon. Oxide on polysilicon is usually at least a three-film structure on a substrate, as shown in Figure 6-7. Typical thicknesses are 1000 Å for the lower oxide, 5000 Å for the

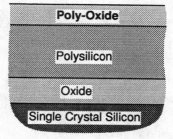

Figure 6-7. The structure involved for the measurement of the thickness of the oxide on polysilicon. As the oxide grows, polysilicon is consumed.

polysilicon, and variable thicknesses for the poly-oxide. As discussed previously, to perform a multiple-film calculation, it is necessary to know accurately the thickness and the index of refraction of each of the underlying films. A further complication is that growth of the oxide causes a decrease in thickness of the polysilicon.

6.8 Simplifications

The author and associates[7,8] have discussed a method whereby the problem can be considered a single-film problem. This involves a specially designed test structure where the thickness of the oxide under the polysilicon is either very near zero or near an integral number of periods. Because the index of refraction of polysilicon is very near that of single crystal silicon, the lower oxide is essentially invisible. It is necessary to determine the value of the effective index of refraction of the composite substrate. Under these conditions, the growth of oxide on polysilicon can be treated as a single-film problem.

The requirement for a special test structure and the need to determine the effective index of refraction for the substrate makes this

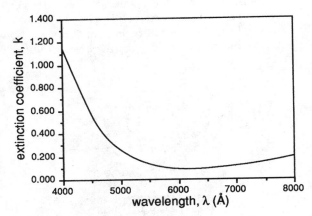

Figure 6-8. Extinction coefficient k as a function of wavelength for polycrystalline silicon, calculated from Cauchy extinction coefficients provided by Mark Keefer, Prometrix Corporation, Santa Clara, CA 95054.

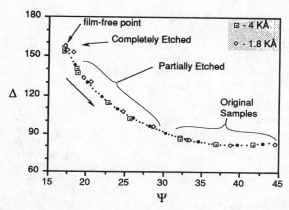

Figure 6-9. Calculated Del/Psi trajectory and measured points for various poly-oxide thicknesses using UV ellipsometry , with λ = 4050 Å. (After Tompkins[9])

method awkward, at best. A simpler solution to this problem[9] is available using single-wavelength ellipsometry if one is willing to use a much shorter wavelength. Figure 6-8 shows a plot of the extinction coefficient for polysilicon. Although the value is quite low for a wavelength of 6328 Å, the value is quite high for a wavelength of 4050 Å. Reference 9 shows that using this wavelength allows polysilicon films thicker than about 1800 Å to be considered substrates, and that we can ignore any underlying materials. Figure 6-9 shows Del/Psi trajectories along with measured values for various oxide thicknesses.

This solution requires either an ellipsometer that has more than one wavelength or an ellipsometer dedicated to this particular measurement. Such ellipsometers have been commercially available and marketed for this purpose since the early 1990s. Kamins[10] has used UV interferometry to measure the thickness of oxide on polysilicon and concludes that a 5000 Å film appears infinitely thick for wavelengths in the range of 2000 to 3500 Å. Reference 9 suggests that this estimate is overly conservative, and that a much thinner film can be treated as a substrate.

6.9   References

1. T. I. Kamins, "Polycrystalline Silicon for Integrated Circuit Applications," Kluwer Academic Publications, Norwell, MA (1988). J. J. Barnes, J. M. DeBlasi, and B. E. Deal, *J. Electrochem. Soc.*, **126**, 1779 (1979).

2. S. Chandrasekhar, A. S. Vengurlekar, S. K. Roy, and V. T. Karulkar, *J. Appl. Phys.*, **63**, 2072 (1988).

3. J. Geist, A. R. Schaefer, J. Song, Y. H. Wang, and E. F. Zalewski, *J. Res. Natl. Inst. Stand. Technol.*, **95** 549 (1990).

4. M. Hirose, M. Taniguchi, and Y. Osaka, *J. Appl. Phys.*, **50**, 377 (1979).

5. R. E. Jones, Jr., and S. P. Wesolowski, *J. Appl. Phys.*, **56**, 1701 (1984).

6. E. A. Irene and D. W. Dong, *J. Electrochem. Soc.*, **129**, 1347 (1982).

7. H. G. Tompkins and B. Vasquez, *J. Electrochem. Soc.*, **137**, 1520 (1990).

8. B. Vasquez, H. G. Tompkins, and R. B. Gregory, *J. Electrochem. Soc.*, **137**, 1523 (1990).

9. H. G. Tompkins, B. Vasquez, T. Mathis, and G. Yetter, *J. Electrochem. Soc.*, **139**, 1772 (1992).

10. T. I. Kamins, *J. Electrochem. Soc.*, **126**, 838 (1979).

Chapter 7
The Effect of Roughness

7.1 General

The development of the basic equations of ellipsometry discussed in Chapter 1 uses the assumption that the interfaces between different materials are plane parallel. In this chapter, we will discuss the effect on the resulting ellipsometry parameters when this condition is not true, i.e., the interfaces are rough. There has been much written about roughness.[1-3] Aspnes *et al.*[3], in a comprehensive article, puts the mathematical treatment in perspective. Much of the material in this chapter was either taken from or suggested by this article.

7.2 Macroscopic Roughness

Roughness can be characterized approximately by a mean height of irregularities about an average plane, and a correlation length be-

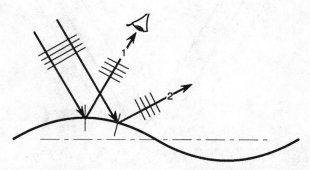

Figure 7-1. Light reflecting from a surface that is macroscopically rough. The dashed line represents the average plane. Ray 1 represents light that goes to the detector whereas ray 2 represents light that is scattered and hence not detected.

tween irregularities.[4] Two different situations arise. When the surface is macroscopically rough, i.e., if the length scale of the irregularities is greater than the wavelength of light, the light is scattered, as suggested in Figure 7-1. If the ellipsometer is aligned properly, the light that reflects from the portions of the sample that are parallel to the average plane will be reflected into the detector as suggested by ray 1 and the light reflected from portions of the sample that are not parallel to the average plane will be directed elsewhere and hence not be detected, as suggested by ray 2.

The effect of macroscopic roughness is simply to reduce the amount of light reaching the detector. If there is sufficient intensity to enable the determination of the parameters by the ellipsometer, it simply measures the optical properties of that part of the surface which is aligned appropriately.

7.3 Microscopic Roughness

We are concerned primarily with microscopic roughness, where the mean height and correlation length of the irregularities are both much less than the wavelength of light. Figure 7-2 suggests such a case. The light interacts with the surface as a whole, rather than interacting

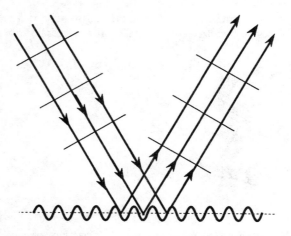

Figure 7-2. Light reflecting from a surface that is microscopically rough. The dotted line represents the average plane.

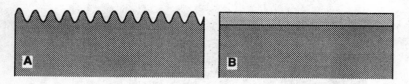

Figure 7-3. (A) A rough surface and (B) the equivalent effective medium on a substrate of the original material.

with each rough spot individually. The effect of the roughness on the far-field radiation pattern can be approximated[5] by one or more layers of an "effective medium," which is sandwiched between a perfect substrate and a perfect ambient. Figure 7-3 illustrates this model. The parameters for this model are the same as for any single film, i.e., the thickness and the complex index of refraction of the "effective" film. Appendix B discusses how to calculate this index.

7.4 Perspective

Although spectroscopic ellipsometry can be used to provide insight about roughness, it is important to understand that, as Aspnes points out, "it is not possible to detect roughness unambiguously from single-wavelength ellipsometric data, where only two parameters are available."[3] Similarly from Raayjmakers and Verkerk,[2] "It must therefore be concluded that single-wavelength . . . ellipsometry cannot be used to measure or even monitor the surface roughness of vacuum-deposited layers." Although we cannot detect roughness with single-wavelength ellipsometry, it is very important to understand how roughness can affect the results of single-wavelength ellipsometry.

By using the plane parallel interface model when roughness is present, incorrect conclusions can be reached. In this chapter, we discuss the effects of roughness on the Del/Psi trajectories.

7.5 Substrate Roughness

We can model the roughness of a surface in several ways. One way is to consider the roughness to be a series of pyramids as shown in

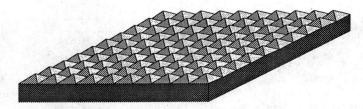

Figure 7-4. Model of rough surface, using pyramids.

Figure 7-4. This might be a reasonable model for films growing from nucleation sites. In this case, the effective medium film is made up of one-third of the substrate material and two-thirds of air. Another example might be the inverse of Figure 7-4 where we have a surface with indentations into the surface. The grain boundaries of a metal might be better represented with this inverted structure of two-thirds material and one-third air. We shall consider the two extremes, where k is near zero and where k is very high.

7.5.1   Near-Dielectric

Let us illustrate the effect of roughness on the Del/Psi values for a material that has a very small value of the extinction coefficient k. For single crystal silicon, $\tilde{N}$ = 3.85 - 0.02 j. If we assume the pyramid model shown in Figure 7-4, the silicon makes up 0.33 of the roughness layer and air makes up 0.67. Using the EMA approximation given in Appendix B, the index for the roughness layer will be $\tilde{N}$ = 1.732 - 0.004 j.

If we were to increase the thickness of the rough layer while retaining the factor of 1/3 silicon and 2/3 air, the Del/Psi trajectory would move as shown in Figure 7-5. In this figure we also plot for reference the trajectory for depositing silicon dioxide. It is clear that it would be difficult to distinguish the presence of roughness from the presence of an oxide film.

Figure 7-6 shows a transmission electron microscope (TEM) image[6] of single crystal silicon which had been etched in a solution intended to etch polysilicon.[7] The original assumption was that this solution did not etch single crystal silicon. Ellipsometer readings gave

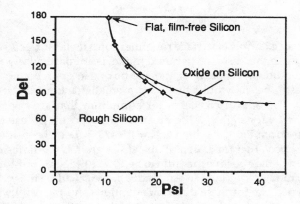

Figure 7-5. Del/Psi values for various thicknesses of a rough surface of silicon. The open diamonds are at 100 Å intervals. Also plotted is the trajectory for SiO$_2$ on silicon. The solid dots are also at 100 Å intervals.

Figure 7-6. TEM image of single crystal silicon surface following a process intended to etch polysilicon but not single crystal silicon.

values of Del/Psi of 137.1°/12.5° and the initial conclusion was that the etch procedure had left an oxide roughly 150 Å thick in place. Auger electron spectroscopy indicated, however, that any oxide present was 5 Å or less in thickness. The TEM image showed that the effect of the etch was to roughen the surface.

7.5.2   Metallic

As an example of surface roughness on a metal surface, we choose aluminum. In this case, the pyramid model is not particularly appropriate. We would expect the roughness to be due to grain boundaries and, in this case, the air would make up a small fraction of the effective medium. The optical constants for aluminum that are quoted in the literature[2,8,9] vary slightly. For illustration purposes, we use the handbook value from Palik,[10] which is $\tilde{N} = 1.5 - 7.6\,j$. Let us consider several cases. Suppose that the indentations at the grain boundaries cause the air fraction of the effective medium to be 10%, 20%, 30%, and 40%. Again using the method outlined in Appendix B, we calculate the index of the effective medium for these different situations and the results are listed in Table 7-1. Also listed are the distances at which the energy density has dropped to $1/e$ of its value at the surface, for the calculated values of k for the effective medium, as discussed in Chapter 1.

Figure 7-7 shows the trajectories for various thicknesses of these films. Since the effective medium is treated as another material (different values of n and k than for the substrate), the trajectories behave like

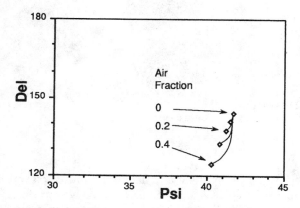

Figure 7-7. Del/Psi trajectories for increasing thickness of rough layers with 10%, 20%, 30%, and 40% air in the effective medium film. The open diamonds represent the Del/Psi location for very thick layers.

TABLE 7-1. Calculated values for n
and k for the effective medium of
various mixtures of Al and air.

Air Fraction	n	k	d (1/e) (Å)
0	1.5	7.6	66
0.1	1.39	6.99	72
0.2	1.26	6.32	80
0.3	1.13	5.56	90
0.4	0.98	4.66	108

those for absorbing films (Section 3.4). The trajectory moves from the film-free position and goes into a very tight spiral around the bulk film position. In this case, the Del/Psi point for very thick films of the effective medium is not that different from the substrate, and the movement of the trajectory is small. Note that although the perspective of Figure 7-7 is the same as a full Del/Psi plot, the abscissa and ordinate of the plot are only one-sixth of the full values.

Considering the depth to 1/e listed in Table 7-1, for roughness greater than a few hundred angstroms, the electromagnetic wave does not reach the bulk metal/rough layer interface because it is absorbed in the rough layer itself. We can then consider the rough layer as a substrate with optical constants as calculated by the EMA. It might be noted that in most cases when the optical constants of aluminum are measured with ellipsometry, the values are usually less than those listed by Palik[10] and different roughness may well be the reason.

From an experimental point of view, Raayjmakers and Verkerk[2] state that "No systematic influence of surface roughness can be observed if the roughness is smaller than ~40 nm" and "the surfaces of these samples may have a different appearance in a scanning electron microscope (SEM)."

7.6   Film Growth with Roughness

7.6.1   Perspective

The common practice for film growth with roughness is simply to take the values given by the instrument as the optical constants of the surface, assume plane parallel interfaces, and go from there. This is to say that the common practice is to ignore the presence of roughness, but to use measured optical constants rather than handbook values.

If one is not going to assume plane parallel interfaces for film growth or film deposition, it is necessary to consider the form of both the substrate/film interface and the film/ambient interface. In film growth, the question arises as to whether the initial roughness remains or whether the roughness decreases. Similarly, the question arises as to whether the film/ambient interface reproduces the substrate/film interface.

For an initial rough surface, the following possibilities seem to exist:

1. Substrate/film roughness remains the same and film/ambient interface is planar.

    Examples of this might be film deposition onto a rough surface by a spin-on method. Chemical vapor deposition methods also often deposit films where the outer surface is planar. In this case, the effective medium approximation is appropriate for dealing with the rough substrate/film interface only. The outside surface would be assumed planar. This would become a two-film problem.

2. Substrate/film roughness decreases and the film/ambient interface is planar.

    Although no literature references are available on this matter, it would seem that this might describe the situation when the substrate material is the moving species during the growth of an amorphous layer of oxide or nitride. In this case, plane parallel interfaces could be assumed after enough film growth had occurred and it is a single-film problem. The value of the index of refraction for the substrate that was measured on the rough surface would be inappropriate, however.

3. The film/ambient reproduces the substrate/film roughness.

    Film growth where the ambient molecules are the moving species might fit this situation. In this case, the effective

medium approximation would be appropriate for both interfaces. This is a three-film problem.

Without extensive ancillary analytical techniques, the roughness
is not known well enough to actually do these calculations. As stated
above, common practice is to treat all of these as single-film problems
and to ignore the roughness.

7.6.2 Estimation of Uncertainty

To understand the uncertainty involved by assuming plane parallel interfaces when roughness is present, let us consider the following
example of oxide growth. Suppose that we have a material with index
$\tilde{N} = 4.0 - 0.2\,j$. This might approximate a form of polycrystalline silicon.
Further suppose that the surface is rough and that the effective layer is
composed of one-third substrate and two-thirds of the overlying material. We consider the growth of a film with index $\tilde{N} = 1.6$, which might
be oxide growth on silicon. Let us consider the third possibility listed
above, i.e., as the oxide grows, the roughness is reproduced. This is
suggested in Figure 7-8 where we show (A) the unoxidized sample and
(B) with 300 Å of oxide. The 300 Å represents the distance from the top
of the substrate roughness to the top of the oxide roughness. We
suppose that the rough layer is 50 Å thick. For the film-free situation,
the effective medium is made of one-third silicon and two-thirds air and
this is a one film situation. The effective index for the rough layer

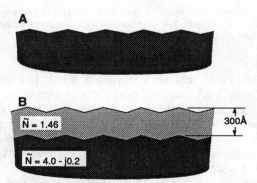

Figure 7-8. (A) Material with rough surface. (B) After growth of 300 Å film, where
the roughness is reproduced.

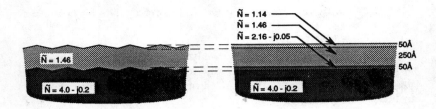

Figure 7-9. Material with 300 Å film with 50 Å of roughness along with effective medium approximation of the same structure.

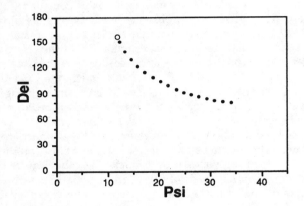

Figure 7-10. Del/Psi points as a function of oxide growth for the structure shown in Figure 7-8 and modeled in Figure 7-9. The open circle is the film-free value and the solid dots are plotted at 50 Å intervals.

is $\tilde{N} = 1.75 - 0.04\,j$ and this is represented by a Del/Psi point of 157.3°/11.9°. When 50 Å oxide is present, the bottom film is composed of one-third silicon and two-thirds oxide. The top film is composed of one-third oxide and two-thirds air. When 100 Å of oxide is present, a three-film situation exists with the same top and bottom films. The center film is oxide (50 Å) with $\tilde{N} = 1.46$. In Figure 7-9 we show the corresponding effective layers and the calculated indices of refraction, in this case for a 300 Å growth of oxide. The Del/Psi trajectory for this kind of film growth is shown in Figure 7-10.

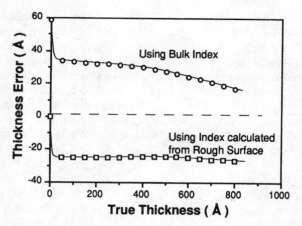

Figure 7-11. Film thickness error caused by making the assumption that we have plane parallel interfaces when in fact both the film/substrate interface and the ambient/film interface are rough (see text).

Let us now suppose that we have a series of rough films as described above with the Del/Psi values shown in Figure 7-10. In real world situations, the state of roughness is usually unknown and some simplifying assumptions are made. Two examples of simplifying assumptions are:

1. Measure the Del/Psi values of the film-free rough surface, use these to calculate an index value for the substrate, and assume that the film growth gives a single film with plane parallel interfaces.

or

2. Use the true index values from the bulk material (determined by some other means) and assume that the film growth gives a single film with plane parallel interfaces.

The trajectory calculated in both these situations falls very near the points shown in Figure 7-10. The primary difference is the starting point (the film-free point). In both cases, let us calculate the trajectory and use it to estimate the thickness of the film. Figure 7-11 shows the error involved in using these simplifying assumptions. For situation #1 listed above, the error is roughly 25 Å for all thicknesses except zero. For situation #2 listed above, the error is 30 to 35 Å for films several hundreds of angstroms. Although Figures 7-10 and 7-11 show thicknesses up to 800 Å, it is probably stretching the imagination to believe that the film would reproduce the surface roughness for films this thick.

The point of Figure 7-11 is that the thickness error involved by either assumption is of the order of or less than the thickness of the roughness. Using a substrate index that was midway between that calculated by the two situations listed above would minimize the error even further.

7.7 References

1. Several examples of articles with title reference to "roughness" and "ellipsometry" are:
   - J. R. Blanco and P. J. McMarr, *Appl. Opt.*, **30**, 3210 (1991),
   - S-M. F. Nee, *Appl. Opt.*, **27**, 2819 (1988),
   - T. V. Vorburger and K. C. Ludema, *Appl. Opt.*, **19**, 561 (1980),
   - K. Brudzewski, *Thin Solid Films*, **61**, 183 (1979),
   - M. J. Verkerk and I. J. M. M. Raaijmakers, *Thin Solid Films*, **124**, 271 (1985),
   - C. F. Fenstermaker and F. L. McCrackin, *Surface Sci.*, **16**, 85 (1969),
   - E. L. Church and J. M. Zavada, *J. Opt. Soc. Am.*, **66**, 1136 (1976),
   - J. P. Marton and E. C. Chang, *J. Appl. Phys.*, **45**, 5008 (1974),
   - T. V. Vorburger and K. C. Ludema, *Appl. Opt.*, **19**, 561 (1980),
   - T. Smith, *Surface Sci.*, **56**, 252 (1976),
   - A. Canillas, J. Campmany, J. L. Andujar, E. Bertran and J. L. Morenza, *Surface Sci.*, **251/252**, 191 (1991),
   - B. J. Thomas, P. Southworth, M. C. Flowers, and R. Greef, *J. Vac. Sci. Technol. B*, **7**, 1325 (1989),
   - I. Ohlidal and F. Lukes, *Opt. Acta*, **19**, 817 (1972),
   - I. Ohlidal, F. Lukes, and K. Navratil, *Surface Sci.*, **45**, 91 (1974).
2. I. J. M. M. Raayjmakers and M. J. Verkerk, *Appl. Opt.*, **25**, 3610 (1986).
3. D. E. Aspnes, J. B. Theeten, and F. Hottier, *Phys. Rev. B*, **20**, 3292 (1979).
4. P. Beckmann, "The Depolarization of Electromagnetic Waves," Golem, Boulder CO, (1968).
5. D. V. Sivukhin, *Zh. Eksp. Teor. Fiz.*, **21**, 367 (1951) and *Zh. Eksp. Teor. Fiz.*, **30**, 374 (1956) [*Sov. Phys. JETP*, **3**, 269 (1956)].
6. TEM photo by Peter Fejes, Motorola, Inc. Mesa AZ 85202.
7. Tom Roche, private communication (1991).
8. A. C. Nyce and L. P. Skolnick, *J. Opt. Soc. Am.*, **65**, 792 (1975).
9. C. J. Del'Oca and P. J. Fleming, *J. Electrochem. Soc.*, **123**, 1487 (1976).
10. "Handbook of Optical Constants," edited by E. D. Palik, Academic Press, New York (1985).

CASE STUDIES

The following are several case studies taken from the re-viewed literature. These cases show various examples of how other workers have used single-angle single-wave-length ellipsometry to study properties of materials. In some cases, *in situ* ellipsometry is used and various examples are shown. In other cases, enhancements were used to measure very small changes in the optical properties. Generally, however, the measurements can be made using standard ellipsometric equipment.

Case 1:
Dissolution and Swelling of Thin Polymer Films

C1.1   General

The swelling and dissolution of thin poly(methyl methacrylate) (PMMA) films has been studied[1-4] by Papanu *et al.* The emphasis of their work was on studying the dissolution of photolithographic resist materials in a developer solution. They studied PMMA, which is a positive-acting electron-beam resist typically developed in ketone/alcohol mixtures. Either swelling or dissolution of the film can result, depending on temperature, solvent composition, and polymer molecular weight.   In the course of several years, the authors studied examples of each of these cases, using various ketone/alcohol mixtures and various processing parameters for the PMMA. Ketones are polar and have minimal hydrogen bonding character, making them good solvents for PMMA. Alcohols are poor solvents for PMMA due to their high degree of hydrogen bonding character. In mixtures, the alcohol moderates the dissolution rate. This allows better process control and enhances contrast.

C1.2   Early Work Using a Psi-Meter

In an early work[1] published in 1984, Flack *et al.* fabricated what they called a "Psi-Meter." This used some but not all of the components of an ellipsometer and the measurement determined Psi only.   The experimental apparatus is shown in Figure C1-1.

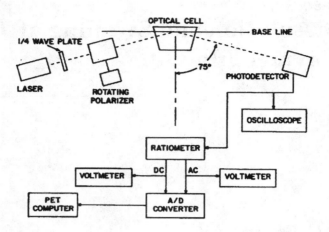

Figure C1-1. Schematic block diagram of the psi-meter setup with the major optical and detection components identified. (After Flack[1])

The sample is immersed in the solvent, so this is an *in situ* measurement. The solvent is either stagnant, agitated, or circulated with a magnetic pump. The temperature of the solvent is held constant at a chosen value.

The polarized laser, quarter-wave plate, rotating analyzer combination gives linearly polarized light with an azimuth angle that varies sinusoidally as a function of time. The ratio of the AC component to the DC component will give a measure of the ellipticity of the polarized light resulting from the reflection. Let η be the ratio of the AC component to the DC component of the photomultiplier output. Zaghloul and Azzam[5] show that $\eta = -\cos(2\Psi)$.

They simply followed this ratio as a function of solution soak time. As the thickness of the film changes due to dissolution, this ratio will also change. Figure C1-2 shows two plots. Endpoint detection is very straightforward since the ratio stops changing.

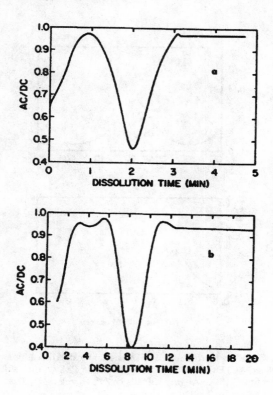

Figure C1-2. Amplitude ratios of the AC to DC components as functions of time for PMMA dissolution in MEK and IPA developer solutions with the following compositions (volume ratios of MEK to IPA): (a) 3:2, (b) 1:1. (After Flack[1])

In a subsequent work[2] published in 1987, they used this instrument to study the dissolution of PMMA in methyl isobutyl ketone (MIBK) as a function of:

- molecular weight of the polymer molecules
- the cooling rate after softbake, and
- post-cool aging of the films.

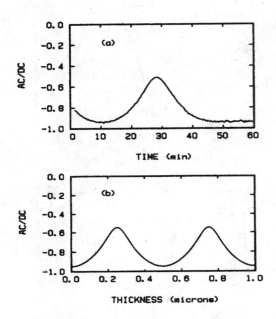

Figure C1-3. (a) Typical experimental AC/DC signal for the dissolution of PMMA in MIBK. (b) Predicted AC/DC vs. film thickness curve for PMMA on a silicon substrate immersed in MIBK. Angle of incidence 75°, wavelength 632.8 nm. (After Manjkow[2])

For this solvent, the primary effect was dissolution. Swelling could be ignored. The expected variation of Ψ (or 2Ψ) as a function of thickness can be calculated and compared to the measured data. Figure C1-3 is one such comparison. The predicted behavior was calculated using McCrackin's program.[6] From the plot, it is reasonably straight-forward to determine the dissolution rate. They found that the first two parameters listed above affected the dissolution rate significantly whereas the third parameter had minimal effect.

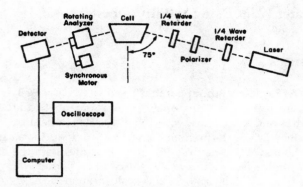

Figure C1-4. Schematic of the rotating analyzer ellipsometer. (After Papanu[3])

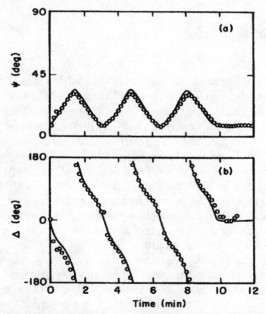

Figure C1-5. Experimental values of (a) Ψ and (b) Δ for dissolution in MIBK at 24.8°C. Solid curves are predicted from an optical model that assumes a linear decrease in film thickness and negligible surface swelling. (After Papanu[3])

C1.3   Later Work Using an Ellipsometer

The Apparatus

A later design[3] of the apparatus is shown in Figure C1-4. In this case, they are using a rotating analyzer ellipsometer and are able to obtain both Δ and Ψ. In this work,[3] they use this apparatus to study solvent systems where the predominant factor is dissolution and solvent systems where swelling occurs. Figure C1-5 shows both the Δ and Ψ data for a sample system similar to that used in the earlier work. In this case, we see the Δ/Ψ values moving cyclically around the Δ/Ψ trajectory until the sample is completely removed. Again, it is quite straightforward to determine dissolution rate.

C1.4   Modeling the Swollen Film

In simple dissolution, we have one film with a changing thickness. When the film swells before dissolution, the problem becomes much more complex. To understand the different contributions, they choose a solvent that causes swelling, but very little dissolution. Adding to the complexity is the fact that there are two different types of swelling. For one type,[7] which they call "Case II" swelling, the interface between the swollen material and the nonswollen material is sharp and the interface propagates into the film at a constant rate. The film can be modeled as two layers with different indices of refraction and changing thicknesses. The progression of swelling simply means that the one layer becomes larger while the other layer becomes smaller. The increase in size due to swelling is also taken into account.

The other type is called "Fickian" swelling. This is a classical case where the solvent diffuses into the polymer under normal diffusion kinetics. The concentration of solvent in the film decreases monotonically as a function of distance from the outer surface and the swelling varies accordingly. To model the film, it is necessary to consider the polymer film to be a series of layers each with a uniform index of refraction. Sullivan et al.[8] discuss the model at length.

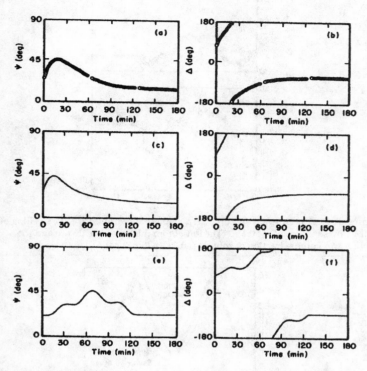

Figure C1-6. Swelling of PMMA in methanol at 19.0°C: (a) and (b) experimental Ψ and Δ data, (c) and (d) predicted values of Ψ and Δ for Fickian swelling, and (e) and (f) predicted data for Case II swelling. The initial film thickness is 1350 nm and the equilibrium volume fraction of solvent is 0.22. (After Papanu[3])

C1.5   Comparison of Data with Model

To illustrate the two different swelling situations, the authors[3] chose two different solvents, each of which would cause swelling with very little dissolution.   Figure C1-6 shows the measured Δ/Ψ data when the solvent was methanol at 19.0°C. The initial film thickness was 13,500 Å.   The final equilibrium volume fraction of the solvent in the swollen film was 0.22. Also shown in Figure C1-6 is the predicted Δ/Ψ

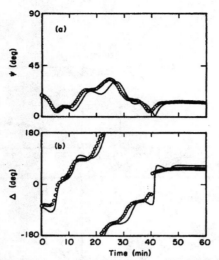

Figure C1-7. Experimental and predicted values of (a) Ψ and (b) Δ for swelling of PMMA in isopropanol at 50.1°C. The initial film thickness is 1726 nm and the equilibrium volume fraction of solvent is 0.39. (After Papanu[3])

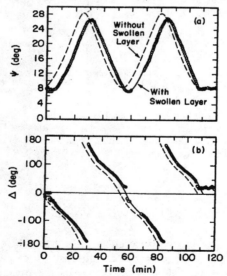

Figure C1-8. Predicted and experimental values of (a) Ψ and (b) Δ for dissolution of monodisperse ($M_n = 7.5 \times 10^5$ g/mol) PMMA in MIBK at 24.8°C. Two predicted curves are shown for dissolution with (solid line) and without (dashed line) a swollen surface layer. (After Papanu[4])

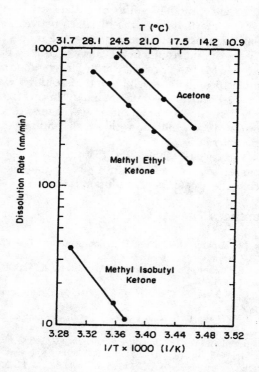

Figure C1-9. Arrhenius plot of the average dissolution rate of polydisperse PMMA in three solvents. The MIBK data are from Ref. 3. (After Papanu[4])

values for both Fickian swelling and for Case II swelling. Clearly, the Fickian swelling matches rather well whereas the Case II swelling does not match. This indicates that for methanol, the concentration of the solvent in the polymer will be varying as a function of the distance from the surface. The polymer thickness increases rapidly at first and then approaches the limit asymptotically.

Figure C1-7 show similar data (the open circles) when the solvent is isopropanol at 50.1°C. Also shown (the solid line) are the predicted values using the Case II model. In this case, the Case II model fits the data rather well. Note that since the interface between swollen and nonswollen polymer moves at a constant rate, the process ends rather abruptly and the endpoint results in the curve becoming flat.

In a study of PMMA with significantly larger molecular weight of the polymer molecules, both swelling and dissolution were found to occur.[4] Figure C1-8 shows the data obtained. In this figure, the data resemble simple dissolution, but an offset was observed. This is modeled by assuming that an initiation time was required during which some swelling occurred.

Finally, Figure C1-9 shows the dissolution rates obtained for several solvents at several temperatures. Note that the smaller the solvent molecule, the greater the dissolution rate.

C1.6   References

1. W. W. Flack, J. S. Papanu, D. W. Hess, D. S. Soong, and A. T. Bell, *J. Electrochem. Soc.*, **131**, 2200 (1984).

2. J. Manjkow, J. S. Papanu, D. W. Hess, D. S. Woane, and A. T. Bell, *J. Electrochem. Soc.*, **134**, 2033 (1987).

3. J. S. Papanu, D. W. Hess, A. T. Bell, and D. S. Soane, *J. Electrochem. Soc.*, **136**, 1195 (1989).

4. J. S. Papanu, E. W. Hess, D. S. Soane, and A. T. Bell, *J. Electrochem. Soc.*, **136**, 3077 (1989).

5. A.-R. M. Zaghloul and R. M. A. Azzam, *Surface Sci.*, **96**, 168 (1980).

6. F. L. McCrackin, *Natl. Bur. Std., Tech. Note 479* (1969).

7. T. A. Alfrey, E. F. Gurnee, and W. G. Lloyd, *J. Polym. Sci. C*, **12**, 249 (1966).

8. M. Sullivan, J. W. Taylor, and C. Babcock, *J. Vac. Sci. Technol. B*, **9**, 3423 (1991).

Case 2:
Ion Beam Interaction with Silicon

C2.1   General

E. A. Irene, D. J. Vitkavage, S. C. Vitkavage, J. L. Buckner, and T. M. Mayer describe in three publications a method of using ellipsometry to study the effect of energetic ion beams on silicon surfaces.[1-3] Plasma-assisted etching techniques are used extensively in integrated circuit processing to obtain highly anisotropic etching. One problem is the creation of a damage layer at the semiconductor surface. This low energy ion bombardment creates a surface alteration which is a few tens of angstroms deep. As an initial approach for development of the method, Irene et al. studied crystalline damage produced by energetic inert gas ion impact.

C2.2   Development of the Analysis Method

In a work[1] published in 1986, Buckner et al. developed the method. They used argon ions ranging in energy from 300 to 1050 eV with current densities of about 100 $\mu A/cm^2$. Silicon <100> wafers were cleaned with the RCA cleaning method[4] and then bombarded with ions with energies of 300, 550, 800, or 1050 eV. They were then aged at room temperature in air for times ranging from several days to 1.5 months.

Analysis was made using ellipsometry and Rutherford backscattering spectrometry. A manual null ellipsometer with a 6328 Å He-Ne laser light source at 70° was used. Data reduction was made using the model shown in Figure C2-1. The indices of refraction for air, silicon dioxide and silicon are the generally accepted values. The index for the

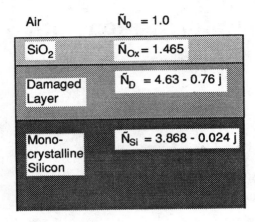

Air $\tilde{N}_0$ = 1.0

SiO$_2$ $\tilde{N}_{Ox}$ = 1.465

Damaged Layer $\tilde{N}_D$ = 4.63 - 0.76 j

Mono-crystalline Silicon $\tilde{N}_{Si}$ = 3.868 - 0.024 j

Figure C2-1. The model used in Refs. 1-3. It consists of a single-crystal silicon substrate covered by a damaged silicon layer of thickness L_D, covered by an amorphous SiO$_2$ film of thickness L_{ox}.

damaged silicon is the value reported by Fried et al.[5] for amorphous silicon created by 20 keV Ar ion implantation into single-crystal silicon.

The inputs to the analysis were the measured Δ/Ψ values and the indices. An iterative method was then used to obtain L_D, the thickness of the damage layer, and L_{ox}, the thickness of the oxide. A real value of L_{ox} was chosen and a solution for L_D obtained. This solution was, in general, complex. The value of L_{ox} was varied to find the value of L_D with the smallest imaginary part and this value of L_D and L_{ox} were then taken to be the solution.

Rutherford backscattering/channeling measurements were used to estimate the damage and oxide layer thickness for comparison with the ellipsometric data. Channeling was used and the damage layer thickness was calculated from the number of displaced Si atoms, assuming that this layer was completely amorphous with an atomic density of 5.0×10^{22} atoms/cm^3. The oxide layer thickness was estimated from the oxygen peak area assuming a stoichiometric oxide with density

2.3 g/cm^3.   Excellent agreement was observed between the damage layer thickness obtained by the ellipsometric method and the RBS method for the original damaged samples and after various anneals.

C2.3   Ion Beam Damage Results

Buckner et al.[1] observed that aging for 1.5 months did not produce a change over aging for a few days. This indicates that the native oxide growth induced by the damage was complete in a few days.

The results of the first work[1] were included in a publication[2] appearing in 1988. In this study, hydrogen, helium, and neon were added as bombarding ions. For argon and neon, again the ion energies varied from 300 to 1050 eV at a dose of 2.5 x 10^{17} ions/cm^2. This was determined to be high enough to ensure total amorphization of the damage layer. For the hydrogen, the energy was held at 1050 eV and the dose was varied from 9.2 x 10^{16} to 1.3 x 10^{18} ions/cm^2. For the helium, the ion energy varied from 250 to 1000 eV and the dose was 7.5 ±1.0 x 10^{17} ions/cm^2.

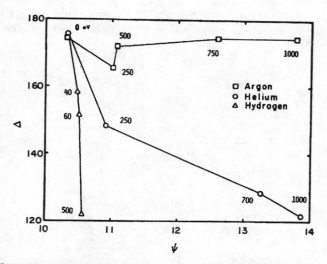

Figure C2-2. Ψ-Δ trajectories as a function of argon, hydrogen, and helium ion energy. (After Buckner[2])

Figure C2-2 shows the results obtained. It is clear that the Δ/Ψ trajectories are sensitive to the identity of the bombarding ion and its incident energy. Each trajectory begins at the Δ/Ψ point, which is characteristic of clean single-crystal silicon (taken to be $175°/10.5°$). To determine the cause for the movement in Δ/Ψ, it is necessary to use the model previously developed.[1] The thickness of the damage layer for the argon bombardment and neon bombardment (not shown in Figure C2-2) obtained with the optical model and by RBS is shown in Table C2-1.

The third and forth entries show the repeatability and spread in the data. The last two entries are for Ne. The agreement is not quite as good as for argon, although still not unreasonable. The greater damage depth with neon is attributed to the greater penetration depth into the substrate with the less massive bombarding ion. It should be emphasized that these thicknesses are for the damage layer under the oxide. Reference 1 indicates that in addition, the oxide film is typically 40 Å thick. This is the part of the damaged layer which was subsequently oxidized due to atmospheric exposure.

Helium behaves somewhat differently from argon and neon. Figure C2-3 shows the helium trajectories for bombardments of 250, 700, and 1000 eV energies. The solid line is the trajectory for growth of oxide on silicon and is shown for reference. The damage trajectory tends to parallel the oxide growth curve. By analogy, this suggests that the major

TABLE C2-1. Damage layer thickness by ellipsometry and RBS. (After Buckner[2])

Treatment	Ellipsometry L_D (Å)	RBS L_D (Å)
Ar 300 eV	20	21
Ar 800 eV	39	- -
Ar 1050 eV	49	49
Ar 1050 eV	52	53
Ne 300 eV	33	23
Ne 1050 eV	79	67

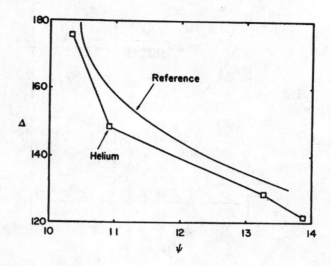

Figure C2-3. Ψ-Δ trajectory under helium bombardment. (After Buckner[2])

effect is the growth of a film whose thickness increases with ion bombardment energy. The RBS gives an oxide thickness of 230 Å and a damage thickness of 170 Å for the highest dose sample.

The optical model used for the heavy ions does not give as good a fit to the RBS data. This is attributed to the fact that for the light ion, the statistical projected range distribution is so broad that the sharp interface assumption is no longer applicable.

Summarizing the damage studies, it appears that for a given ion, higher energy gives thicker damage layers. It also appears that lighter bombarding ions give thicker damage layers.

C2.4   Damage Removal

In the Ref. 1, the authors concluded that annealing in N_2 returned the Δ/Ψ values to near that of the original samples, i.e., the annealing removed the damage. In a third work[3] also published in 1988, the same

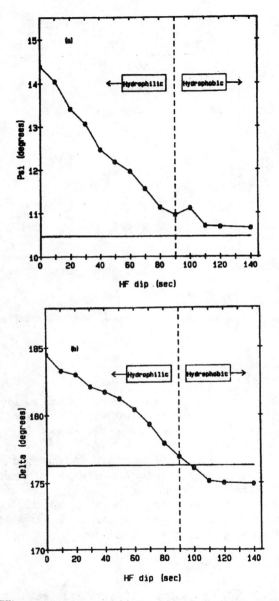

Figure C2-4. Ellipsometric parameters (a) Ψ and (b) Δ for both 3.5 keV Ar ion beam silicon (open circles) and crystalline (nonbombarded) silicon (solid line) after repeated HF dipping. (After Vitkavage[3])

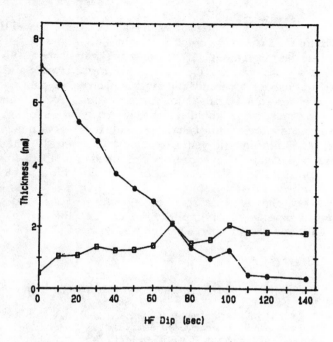

Figure C2-5. Measured oxide layer thickness L_{ox} (open squares) and damaged silicon layer thickness L_D (filled circles) for 3.5-keV Ar ion beam bombarded silicon after a series of 10 s HF dips. (After Vitkavage[3])

optical model is used to measure the removal of the damage by wet etching.

In this work, the samples were damaged with argon ions with 3.5 keV energy. The damage thickness was greater than in the previous study, as would be expected. The oxide thickness was somewhat thinner. In this case, the damage layer thickness was about 70 Å and the oxide thickness was about 10 Å.

The damage was removed by a series of 10 s dips in concentrated HF to remove some of the silicon surface. Following each dip, the samples were rinsed in deionized water and dried. Several Δ/Ψ measurements were then made across the wafer, from which L_D and L_{ox} were determined.

Figure C2-4 shows the values of Δ and Ψ after repeated dipping and Figure C2-5 shows the calculated values of L_D and L_{ox}, using the previously described model. L_{ox} varies only slightly between 10 and 15 Å (1 and 1.5 nm) whereas L_D is seen to decrease rapidly and nearly linearly until about 90 seconds dip time. A change in wetting behavior was also observed to occur at about the same time. For etch times less than 90 seconds, water would wet the surface, whereas for etch times over 90 seconds, non-wetting behavior was observed. For a nondamaged surface, the ellipsometric values remained unchanged. The rate of removal of the damage layer can be determined to be about 40 Å/min. This is considerably faster than that of crystalline silicon, which is reported[6] to be less than 1 Å/min.

X-ray photoelectron spectroscopy analysis observed a small amount of expected argon impurity immediately after bombardment. After 140 seconds of etching, no argon was detected.

C2.5   References

1.   J. L. Buckner, D. J. Vitkavage, E. A. Irene, and T. M. Mayer, *J. Electrochem. Soc.*, **133**, 1730 (1986).
2.   J. L. Buckner, D. J. Vitkavage, and E. A. Irene, *J. Appl. Phys.*, **63**, 5288 (1988).
3.   S. C. Vitkavage and E. A. Irene, *J. Appl. Phys.*, **64**, 1983 (1988).
4.   W. Kern and D. A. Puotinen, *RCA Rev.*, **31**, 187 (1970).
5.   M. Fried, T. Lohner, E. Jaroli, Gy. Vizkelethy, G. Mezey, J. Gyulai, M. Somogyi, and H. Kerkow, *Thin Solid Films*, **116**, 191 (1984).
6.   R. Singh, S. J. Fonash, S. Ashok, P. J. Caplan, J. Shappirio, M. Hage-Ali, and J. Ponpon, *J. Vac. Sci. Technol. A*, **1**, 334 (1983).

Case 3:
Dry Oxidation of Metals

C3.1 General

In many cases, the oxidation of metals can be readily measured with ellipsometry. The requirements are the same as for all ellipsometry. A reasonable approximation to plane parallel interfaces is required. The metal must be smooth enough that enough light returns to the ellipsometer after the reflection. It is much more straightforward if the metal is thick enough that no light penetrates to the underlying medium.

The optical constants of the metal depend on properties such as grain size, etc. Since these properties vary depending on fabrication conditions, it is imperative that the optical constants of the metal be measured for each different condition.

Some of the author's work in this area is described in Chapter 4. In this case study, we present some oxidation studies by others. Because of the nature of the oxidation, we separate thermal and plasma oxidation from electrochemical oxidation.

C3.2 Oxidation of Bismuth at Room Temperature

Metal Surface Roughness

In 1985, Atkinson and Curran published a study of the oxidation of vacuum-deposited bismuth films.[1] It had been shown previously[2] that the optical properties of thin films of this material depend strongly on the microscopic surface roughness. More recent work[3,4] had shown that the roughness of the film could be controlled by controlling the substrate temperature during deposition. In the 1985 work,[1] it was determined that the optimum substrate temperature for depositing smooth films was 353 K and this temperature was used for depositing the films to be used in the oxidation studies.

Experimental Details

Bismuth films about 500 Å thick were deposited onto borosilicate glass substrates. Ellipsometric measurements were made immediately after deposition and at intervals of 24 hours for a period of 9 days. The samples were stored at room temperature in a dust-free environment in air at atmospheric pressure. Optical measurements were made at six wavelengths in the visible region of the spectrum at an incident angle of 45°.

Results

Valley[3] made a measurement at the metal substrate interface as shown in Figure C3-1. Since this interface is free from the effects of oxidation, the aging time after deposition before measurement was unimportant and the optical constants can be readily determined. Atkinson states[1] that the optical constants obtained in this way are in close agreement with those obtained from direct measurements made with conventional ellipsometry if the measurement is made within 30 minutes after removal from the vacuum system.

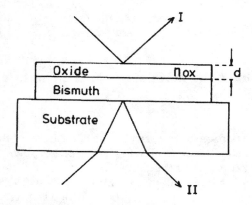

Figure C3-1. Optical model for an oxidized bismuth film: I, air-film interface measurement; II, film-substrate measurement. (After Atkinson[1])

Figure C3-2 show the optical constants obtained for the freshly made bismuth films. In this figure n' represents the real part of the index and n" represents the extinction coefficient which in this book is commonly denoted as k. One question that might arise is whether a 500 Å film is thick enough to be considered bulk (light does not penetrate to the underlying medium). With these values of the extinction coefficient, one can calculate using Equation 8 in Chapter 1 that the 1/e thickness is about 125 Å. The 500 Å film is four times thicker than this, thus qualifying as bulk.

The growth of the oxide film is shown in Figure C3-3 where the oxide has grown to a thickness of 16 Å after 9 days in air. Also shown are the values obtained for the index for the film. It should be noted that the precision with which both the index and the thickness can be obtained simultaneously is not good. It also might be noted that there is an assumption that the extinction coefficient for the oxide is assumed to be zero. The data for this figure were obtained using a wavelength of 4921 Å. The authors observe that the data can be fit to either a logarithmic or a parabolic growth law.

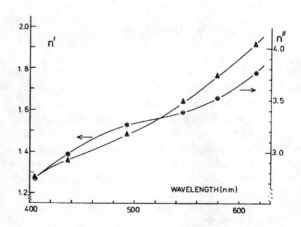

Figure C3-2. Dispersion of the optical constants of a freshly made bismuth film. (After Atkinson[1]) Note that the real and imaginary parts of the index of refraction (denoted as n and k in this text) are denoted as n' and n" in this figure.

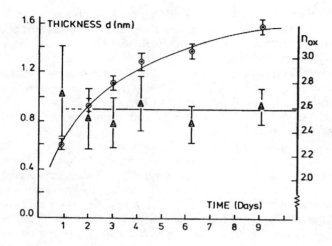

Figure C3-3. Temporal variations in the thickness (circles) and the refractive index (triangles) of the oxide layer. (After Atkinson[1]) Note that $\lambda = 4921$ Å.

C3.3 Plasma Oxidation of Tantalum

Sample Preparation

In 1974, Leslie and Knorr reported[5] a study using an early version of a rotating element ellipsometer. The tantalum was oxidized in an oxygen glow discharge and the oxide growth was observed *in situ*. The sample was a piece of 0.1-mm-thick capacitor-grade Ta sheet. The cleaning procedure was to heat the material to a white heat in a vacuum of 10^{-7} Torr by passing a large current through the sample. This was further cleaned after mounting in the oxidation chamber by an argon glow discharge.

Oxidation and Measurement

The measurement apparatus is shown schematically in Figure C3-4. The angle of incidence was 59.5° and the values of the null position of the polarizer and analyzer were obtained once every 5 s.

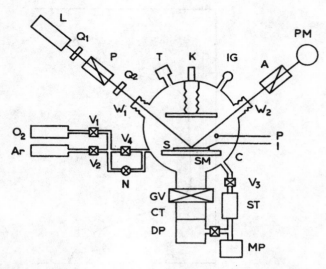

Figure C3-4. Schematic diagram of the experimental arrangement. The components are: laser (L), two quarter-wave plates (Q_1, Q_2), polarizer (P), analyzer (A), photomultiplier (PM), sample (S), sample mount (SM), windows (W), drive current wire (I), voltage probe (P), vacuum chamber (C), valves (V), needle valve (N), gate valve (GV), sorption trap (ST), liquid nitrogen cold trap (CT), mechanical pump (MP), diffusion pump (DP), thermocouple gauge (T), ionization gauge (IG), and the glow discharge cathode (K). (After Leslie[5])

Following the glow discharge cleaning of the sample in argon, the system was flushed with flowing oxygen. By dynamic pumping, the oxygen pressure was stabilized at about 55 mTorr and the oxygen discharge started. A cathode voltage of -900 V and a current of 10 mA were used for the discharge. Ellipsometer readings were taken for a period of 15,000 s, during the growth of 1500 Å of oxide.

The null values of the polarizer and analyzer were converted to Δ/Ψ points and the trajectory obtained is shown in Figure C3-5. They found that the data could be modeled much better if they assumed that a 20 Å oxide were present before the plasma oxidation began. They obtained a value of the index for the substrate of $\tilde{N}_s = 2.3 - 2.6\,j$. The value determined for the oxide film was $\tilde{N}_f = 2.21 - 0.0\,j$. In comparing

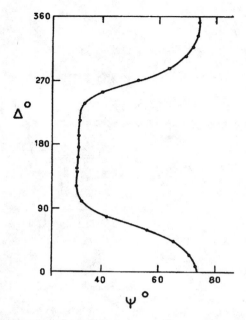

Figure C3-5. Ellipsometric results for the plasma oxidation of Ta. The solid curve represents the experimental results and the open circles denote the theoretical fit obtained using a refractive index of 2.21 for the Ta oxide. Experimental conditions: sample area, 162 mm^2; constant drive current, 10 mA; oxygen pressure, 0.055 Torr; cathode voltage, -900 V; wavelength, 6328 Å; and angle of incidence, 59.5°. (After Leslie[5])

their work with a previous work,[6] Leslie and Knorr point out that having more Δ/Ψ values in the region near $\Delta = 0$ significantly increased the accuracy of the resulting value for the film index. This is the "sweet spot" mentioned in Chapter 4.

C3.4 Thermal Oxidation of Nickel

Introduction and Experimental Setup

In 1974, a study was reported[6] of the kinetics of formation of thin oxide films on the surface of nickel. The metal samples were metallurgically polished and washed with water, acetone, and methanol, and this was followed by drying in an argon stream. The oxidation was done in a quartz tube open to the atmosphere. Temperatures of 100 to 300°C were used for times ranging from 0.5 to 18 h. The ellipsometric measurements were made with a manual null instrument with wavelength of 5461 Å and an angle of incidence of 60°.

Results

The index of refraction for the substrate was determined to be about $\tilde{N}_s = 1.8 - 1.9\,j$. Since the thickness of the initial air-formed film was about 10 Å, the values reported are thickness increases. For samples oxidized in the range of 100 to 250°C, the index of refraction for the oxide film was determined to be about $\tilde{N}_f = 2.9 - 0.09\,j$, where the uncertainty

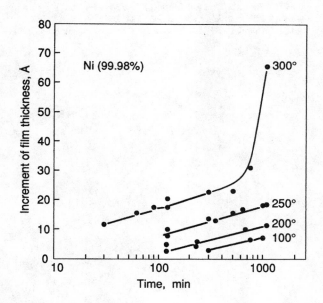

Figure C3-6. Dependence between increment of film thickness and time of oxidation. (Data from Szklarska-Smialowska[6])

in n was ±0.1 and the uncertainty in k was ±0.02. For films oxidized in the range of 250 to 300°C, with thicknesses greater than about 25 Å, a change was observed in the values of n and k. In this range, the index of the film was $\tilde{N}_f = 2.6 - 0.16$ j, with the same uncertainty. Figure C3-6 shows the thicknesses versus time for the different temperatures. The conclusion drawn[6] was that the oxidation occurs according to a logarithmic law for oxidations below 250°C and according to a parabolic law above 250°C.

C3.5 References

1. R. Atkinson and E. Curran, *Thin Solid Films*, **128**, 333 (1985).
2. R. Atkinson and P. H. Lissberger, *Thin Solid Films*, **17**, 207 (1973); A. Barna, P. B.Barna, R. Fedorowich, G. Radnoczi, and H. Sugawara, *Thin Solid Films*, **36**, 75 (1976).
3. L. Vallely, PhD. Thesis, The Queen's University of Belfast (1981).
4. Y. Namba and T. Mori, *J. Appl. Phys.*, **46**, 1159 (1975); K. Abdelmoula, B. Pardo, C. Pariset, and D. Renaro, *Thin Solid Films*, **62**, 273 (1979).
5. J. D. Leslie and K Knorr, *J. Electrochem. Soc.*, **121**, 263 (1974).
6. Z. Szklarska-Smialowska and H. Oranowska, *Corrosion Science*, **16**, 355 (1974).

Case 4:
Optical Properties of Sputtered Chromium Suboxide Thin Films

C4.1   General

Photomasks used in the integrated circuits industry typically use layers of Cr/Cr_xO_y on glass. In the preparation of the chromium, oxygen is often introduced to alter the film complex index of refraction, specifically to lower the reflectance. In 1984, Hsia[1] published a study of the optical properties of several types of Cr/Cr_xO_y films. Auger, ellipsometry, and reflectivity measurements were also reported.

C4.2   Film Preparation and Auger Analysis

Three types of Cr/Cr_xO_y films with oxygen content described as poor, medium, and rich were deposited. These are referred to as type A, B, and C, respectively. The films were deposited onto borosilicate glass. Type A was prepared by sputtering 700 Å of pure chromium onto the glass substrate in a helium environment. Adventitious oxygen in the sputtering environment was incorporated into the film and the resultant atomic ratio of Cr to O was determined by Auger analysis to be about 4:1.

Type B films were produced by intentionally introducing oxygen into the sputtering chamber during deposition. The Type B films were 1250 Å thick and the atomic ratio of Cr to O was measured to be 3:2.

In both Type A and Type B, the oxygen distribution is reasonably uniform throughout the film. To further increase the oxygen concentration while maintaining high optical density, Type C films were prepared in a layer structure. The first step was to prepare a film similar to Type A. The oxygen content in the chamber was significantly increased while

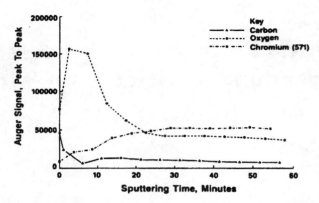

Figure C4-1. Auger profiles of Type C chromium film. Thickness = 1900 Å. (After Hsia[1])

continuing to deposit Cr. The Auger profile of the resultant film is shown in Figure C4-1. The total thickness is about 1900 Å while the thickness of the oxygen-rich layer is about 500 Å.

C4.3   Optical Measurements

Ellipsometric analysis was done with a rotating element ellipsometer with the light having $\lambda = 4880$ Å. Table C4-1 gives the measured Δ/Ψ values for the three different type films. The thickness is also listed. For films of types A and B, the Auger analysis indicates that the assump-

TABLE C4-1.   Indices of refraction calculated from ellipsometry data Ψ and Δ. Thickness measured independently. (After Hsia[1])

Type	Ψ	Δ	n	k	Thickness (Å)
A	29.12	135.18	3.47	3.83	700
B	19.42	109.96	2.67	1.86	1250
C	24.23	52.03	1.39	0.88	1900

tion of a homogeneous material is reasonable. Treating the material as bulk, they calculated the optical constants as discussed in Chapter 3. The results are also listed in Table C4-1.

By using Equation 8 in Chapter 1 and the values of k for films of type A and B, one concludes that the depth that the intensity drops to $1/e$ of its surface value is about 100 and 210 Å, respectively. Clearly the assumption of a bulk material is reasonable since the thicknesses of the material are significantly greater than three times these values. For the case of Type C material, they also made the same assumption (bulk material). If we calculate the depth where the intensity has dropped to $1/e$ of the surface value, we obtain a depth of about 440 Å. Since the oxygen-rich part of this film is only about 500 Å, the ellipsometer is not simply sensing this top layer. The values of Δ/Ψ values listed in Table C4-1 for film Type C are then the optical parameters of the composite film and, therefore, dependent on the thickness of the oxygen-rich layer. Although they list values for n and k, these values will not be representative of either the oxygen-rich part or the lower layer of the Type C film.

Figure C4-2 shows the reflectance of the different type films as a function of wavelength. Although the reflectance varies from 60% for the oxygen-poor film to 10% for the oxygen-rich layered film, the reflectivity remains relatively flat in the range of interest for most photolithography.

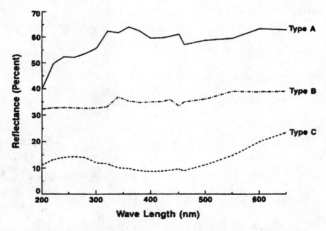

Figure C4-2. Reflectance as a function of wavelength. (After Hsia[1])

TABLE   C4-2. Comparison of reflectance between the direct measurement and calculation (%). (After Hsia[1])

Type	Measured	Calculated
A	58.1	57.1
B	34.6	40.0
C	8.6	13.4

Table C4-2 gives the comparison between the measured reflectance (presumably at a wavelength of 4880 Å) and the calculated reflectance using the n and k values in Table C4-1 and Equation 15 in Chapter 1. The comparison is rather good for types A and B. Although the values for Type C films are not as close, considering the fact that the listed n and k values are for a composite film, the comparison is not unreasonable.

C4.4   Reference

1. L. C. Hsia, *J. Electrochem. Soc.*, **131**, 2133 (1984).

Case 5:
Ion-Assisted Film Growth of Zirconium Dioxide

C5.1 Experimental Apparatus

Martin *et al.*[1] have studied the deposition of zirconium dioxide films onto glass substrates. The films are deposited by electron beam evaporation (E-Beam). Film adhesion and film density are improved if the condensing material is bombarded with ions during deposition. The experimental apparatus[2] is shown in Figure C5-1. The ellipsometric measurements are made at a wavelength of 6328 Å and an angle of incidence of 70°.

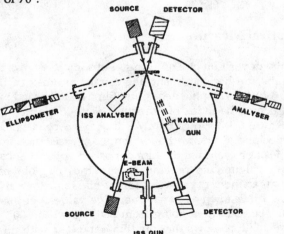

Figure C5-1. Diagram of the UHV deposition and analysis system. (After Martin[1])

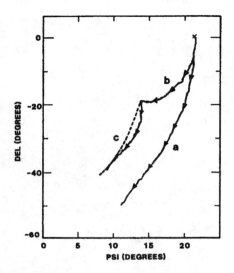

Figure C5-2. Ellipsometer (Del-Psi) traces of zirconium dioxide films deposited by (a) 600 eV oxygen-ion-assistance; (b) evaporation followed by ion-assisted deposition; (c) theoretical curve of a film of refractive index 2.13 deposited on the evaporated layer of index 1.84. (After Martin[1])

C5.2   Optical Measurements

For the deposition of ZrO_2 films, ellipsometry shows that in the absence of the ion bombardment, the refractive index of the film is typically $\tilde{N} = 1.84$. When the depositing film is bombarded by a 600 eV oxygen ion beam, the index of refraction of the deposited material is $\tilde{N} = 2.13$. Figure C5-2 shows the Δ/Ψ trajectory for (a) an ion beam-assisted deposition, (b) deposition without ion assistance, and (c) ion-assisted deposition on top of the film deposited without ion assistance.

From the figure, the trajectory starts at the Δ/Ψ point of $0°/21.2°$. This represents an index of refraction for the glass substrate of $\tilde{N} = 1.475$. In the figure, Δ is shown as having negative values. In this text, this graph would be shown as going from 360° down to 300°. A quick trajectory calculation shows that the film deposited with no ion assistance is roughly 400 Å thick.

The authors find that the change from low to high index material occurs over the first few ion-assisted monolayers. Ion-scattering measurements reveal that the surface density of zirconium atoms is increased about 20% by the ion bombardment. This is consistent with the concept of ion beam densification leading to high refractive indices for ion-assisted films.[3]

C5.3 References

1. P. J. Martin, R. P. Netterfield, W. G. Sainty, and C. G. Pacey, *J. Vac. Sci. Technol. A*, 4, 463 (1986).
2. R. P. Netterfield, P. J. Martin, W. G. Sainty, C. G. Pacey, and R. Duffy, *Rev. Sci. Instrum.*, 56, 1995 (1985).
3. P. J. Martin, R. P. Netterfield, and W. G. Sainty, *J. Appl. Phys.*, 55, 235 (1984).

Case 6:
Electrochemical / Ellipsometric Studies of Oxides on Metals

C6.1 General

Ellipsometry has been used extensively to study various electro-chemical processes. To illustrate the studies of oxides on metals, we shall use some of the very extensive works of DeSmet and Ord (with a variety of coworkers)[1-25, 50] as examples. The references listed provide a sampling of their work done at the University of Alabama and at the University of Waterloo in Ontario, Canada. There are various other examples[26-40] of electrochemical studies using ellipsometry, but the work of DeSmet and Ord will suffice to illustrate the methods. We shall include the growth of anodic oxides on metals and the deposition of metal oxides from solution.

C6.2 Experimental Methods

The basic electrochemical cell is shown schematically in Figure C6-1. The sample where the film is to be grown or deposited and a counter-electrode are immersed in an electrolyte and the power supply drives a chosen constant current between them. In these experiments, most of the sample was masked, exposing a fixed area. As the film grows or is deposited, more voltage is required to continue the process. The reference electrode measures the potential difference between the electrolyte and the substrate (the potential drop across the growing film). The electrolyte is often bubbled with argon or nitrogen before and

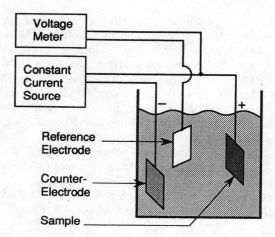

Figure C6-1. Basic electrochemical setup shown schematically. The setup shown is for anodic processes. For cathodic processes, the polarity of the current source is reversed.

during the experiment. In the various studies described here, the counter-electrode was often platinum and the reference electrode was either platinum or a mercury-mercurous sulfate electrode.

One of the requirements when electrochemistry is done along with *in situ* ellipsometry is that the light beam strike the sample and be

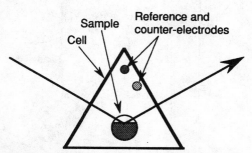

Figure C6-2. One example of a sample cell shown schematically. This cell representation is drawn from descriptions in the work of DeSmet, Ord, and coworkers.[1-25] The sample represents an end-on view of a cylinder with an optical flat ground on one side.

reflected into the ellipsometer. It is also important that the light enter
and leave the electrolyte perpendicular to the electrolyte surface. One
possible sample setup is shown in Figure C6-2. In most of the work of
DeSmet, Ord, *et al.*, for the growth of films, the electrode was a single
crystal rod of the material of interest. For deposited films, the electrode
(and sample substrate) was a platinum single crystal rod. In both cases,
a flat region was ground on the rod for the reflection. In Figure C6-2, we
suggest this by showing the rod end-on. The additional sample preparation will be described later in the respective sections. The electrolyte
was enclosed in a hollow equilateral prism, which held about 60 ml of
electrolyte. The angle of incidence of the light was 60° so that the light
entered and left the electrolyte normal to the surface. Although the
positioning of the reference and counter-electrode is unknown, we acknowledge their presence in the figure. A second example of another

Figure C6-3. Second example of an ellipsometric-electrochemical cell. (After Lukac[26])

worker's ellipsometer cell[26] is shown in Figure C6-3. Note that, in the ellipsometry equations, the ambient medium will be the electrolyte rather than air, as has been the case previously.

Several ellipsometers were developed over the years by DeSmet, Ord, and coworkers and each one evolved from the previous version. They were basically null instruments with a 60° angle of incidence and a wavelength of 6328 Å. The polarizer and analyzer were driven by stepper motors and controlled by computer and were able to make measurements about once per second. This is often enough for the resulting plots to appear as solid lines. It should be pointed out that many processes occur slow enough that it is not necessary to take data at this high speed.

Rather than plot Δ and Ψ, the data in most of their works are plotted in terms of the position of the polarizer, P, and the position of the analyzer, A. The quarter-wave plate is set for retardation and the orientation with its fast axis at 45° to the plane of incidence[1] (QWP = -45°). The resulting relationships between Δ/Ψ and P/A are $\Delta = 90° + 2P$ and $\Psi = A$.

C6.3   Oxide Growth: 1. Zirconium

Oxide Growth in General

In electrochemical oxide film growth, the external constant current power supply causes a potential to be built up across the growing film and this is the driving force to cause either metal ions from the substrate or oxygen ions from the electrolyte to traverse the film. The total charge per unit area transferred (current density multiplied by the time) can be converted to mass added per unit area. This along with thickness determined by ellipsometry can be used to estimate film density.

Experimental

For the anodic growth of an oxide film on zirconium, Hopper, Wright, and DeSmet[9] used an aqueous solution of 0.1M Na_2CO_3 as an electrolyte. All measurements were made with the electrolyte at room temperature. The sample was a high purity (99.99%) zirconium rod, 0.6 cm in diameter with a flat ground on one side for the optical experiments, as suggested by Figure C6-2. The sample was further prepared by

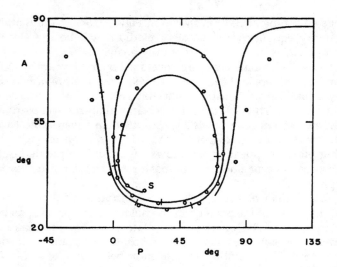

Figure C6-4. Ellipsometric data observed on growth of an anodic film on zirconium in 0.1M Na_2CO_3 at 0.5 mA/cm^2. Data start at point S and trace out loops clockwise. The solid curve represents the best theoretical fit to the data, obtained with a film of refractive index 2.17 - 0.033 j on a substrate of index 2.21 - 3.04 j. Hatch marks indicate thickness increments of 500 Å. (After Hopper[9])

removing all visible signs of oxide with fine emery paper and then applying a chemical polish described by Cain.[41]

Results

The anodization was done with a steady-state current density of 0.5 mA/cm^2 and the ellipsometric trajectory is shown in Figure C6-4. The open circles are a few of the measured values and the solid line is the best theoretical fit to the data. The outward spiral, for a P/A curve like a Δ/Ψ curve, indicates the growth of a film with a nonzero value of k. The best fit was obtained with a refractive index $\tilde{N}_f$ = 2.17 - 0.033 j for the film and $\tilde{N}_s$ = 2.21 - 3.04 j for the substrate. The fit is quite good up to thicknesses of about 44,000 Å. They[9] attribute the deviation at greater thicknesses to slight nonuniformities in optical properties of the film or the substrate.

They observe a linear thickness-potential relationship. This implies that at the constant current chosen, the electrical field in the growing film remained constant.   By using other choices of current density, they determined that the electric field in the film varies directly with the logarithm of the current density.   The optical properties of the film, however, did not depend on the applied current density (or the electric field in the film). This is in contrast to anodic films grown on tantalum, niobium and tungsten[4] where the optical properties were dependent on the electric field.

C6.4   Oxide Growth: 2. Titanium

Experimental

As a second example of the growth of an anodic oxide electrochemically, let us consider the oxidation of titanium, studied by Ord, DeSmet, and Beckstead.[22] Again, the sample was a cylinder with an

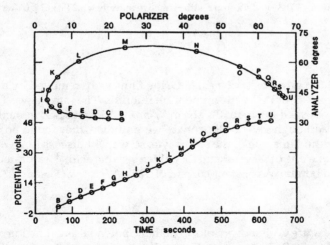

Figure C6-5. Anodic oxidation of titanium in acetone-based electrolyte at a constant current density of 325 µA/cm^2. Corresponding points at 30 s intervals are identified on the potential-time plot in the lower portion of the figure and on the optical data plot in the upper portion. (After Ord[22])

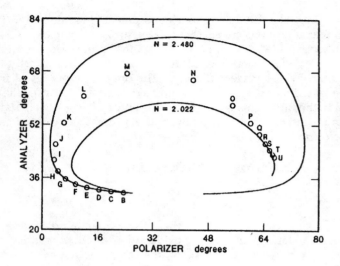

Figure C6-6. Analysis of the optical data from Figure C6-5. Theoretic al curves are shown for the growth of transparent films with the indicated indices to a maximum thickness of 1200 Å on a substrate with refractive index $\tilde{N} = 3.23 - 3.62$ j. (After Ord[22])

optical flat ground on one side. Oxide films were removed by immersion in HF saturated with ammonium fluoride. The sample was further electropolished for about 8 s at 3 mA/cm^2 in 12.5% H_2SO_4 in methanol.

Although several electrolytes were studied, they found no difference in the films due to the electrolyte, so we will discuss films formed in acetone with 1% water and 22 g/liter benzoic acid,[42] saturated with sodium borate. A constant current of 325 μA/cm^2 was used.

Results

Figure C6-5 shows a plot of both the potential as a function of time as well as the P/A trajectory. Data points at 30 s intervals are marked on the lower curve with corresponding marks on the upper curve. Whereas in the previous example, the potential/time curve was linear, in this case, the slope is constant up to about 25 V, increases to a maximum between 30 and 35 V, and then decreases progressively beyond 40 V. This change in slope suggests a change in film structure.

Figure C6-6 shows the P/A data along with two calculated P/A trajectories. Since titanium oxidizes immediately when exposed to air, it is expected that even at time zero, a thin oxide film exists. They used a technique similar to those discussed in Chapter 4 to determine the film-free point. They used the requirement that a plot of film thickness vs. potential extrapolate to zero thickness at the zero of overpotential determined by the open-circuit transient analysis (described in Ref. 22). The resulting value was determined to be $\tilde{N}_s = 3.23 - 3.62$ j. This agrees favorably with values reported in the literature.[43]

The points from B to G can be fitted with a curve that represents the growth of a transparent film with $\tilde{N}_f = 2.48$. In going from point B to point G, the film thickness increases from about 30 Å to about 270 Å. In the lower curve in Figure C6-5, the potential/time curve is quite linear from points B to G. The authors conclude that in this regime, the film formed is "clearly a compact film with a density comparable to that of the crystalline forms of the oxide."

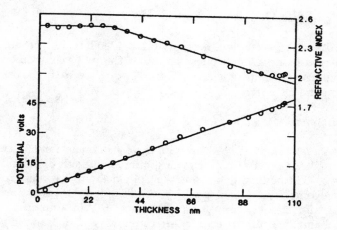

Figure C6-7. The dependence of potential and refractive index on film thickness as determined by the optical analysis in Figure C6-6. (After Ord[22])

The progressive deviation from the $\tilde{N}_f = 2.48$ growth curve at the higher anodization potentials (greater thickness) indicates that there is a change in the structure of the oxide film. Additional assumptions are required for this regime. The direction of deviation is opposite to that of an absorbing film ($k > 0$), as in the previous example. The deviation can only be rationalized as a decrease in refractive index. One could either assume that the changed part is the new growth or one could assume that the entire film is changed. The authors assumed that the entire film changed structure, which simplifies the analysis. The second theoretical curve plotted in Figure C6-6 is for $\tilde{N}_f = 2.022$. This is the value required for the curve to pass through point S. The two calculated curves bracket all of the points and define a range of values for the film index.

The film index value that causes the curve to pass through each point can be calculated (using the homogenous model) and Figure C6-7 shows this value plotted versus the film thickness. In addition, this figure shows the potential versus thickness for each point. The decrease in refractive index begins at a thickness of 310 Å (31 nm) and continues to breakdown at 1030 Å (103 nm). The relationship between potential and thickness appears to be linear over the entire range of thickness indicating a constant electric field.

C6.5   Oxide Growth: 3. Vanadium

As a third example of the growth of oxide films electrochemically, we discuss oxide growth on vanadium. This will also demonstrate another aspect of ellipsometric analysis. For this, we use two works[7,8] of Clayton and DeSmet.

Experimental

Vanadium rods, 99.98% pure, 0.25 in. in diameter, and 0.7 cm long were used. Again an optical flat was ground on the side of the rod. This area was polished using a stainless steel block and various grades of lapping compound, finishing with jeweler's rouge. It was then electropolished for about 2 min at a potential of 16 V in a mixture of 75% ethanol and 25% sulfuric acid.[44] After mounting in the sample holder, the sample was washed in distilled water, dried, anodically oxidized to a potential of 50 V and rinsed again in water (the oxide of vanadium is soluble in water) to remove the oxide. This procedure was repeated several times to ensure reproducibility.

The electrolyte used was similar to that first reported by Keil and Salomon[45] and appears to be the only electrolyte that will sustain the formation of a vanadium oxide film. The electrolyte contained 19 g of $Na_2B_4O_7 \cdot 10H_2O$, 10.6 ml of water, and enough acetic acid to make 1.0 liter of solution. The solution was bubbled with argon both prior to and during the experiments. The index of refraction of this solution is 1.3707 at a wavelength of 6328 Å.

Vanadium differs from other metals described previously in that there is a transition region that occurs prior to the commencement of oxide growth and in this transition period, vanadium ions go into solution.[46] Growth of the anodic oxide could not be initiated using the current densities in which the authors were interested. A current density of 1.5 mA/cm² was used to initiate growth. This was reduced to 275 µA/cm² upon obtaining steady-state growth.

Results

Figure C6-8 shows some of the P/A points for steady-state growth with a current density of 275 µA/cm². More than 1200 data points were recorded during the experiment, hence the points shown are representative points. The start point is simply the point measured before any current was applied. The P/A point traces out a counter-clockwise trajectory and closes on itself approximately. The second loop, represented by the squares, does not exactly coincide with the first loop, but in the first paper,[7] Clayton and DeSmet ignore this.

The data can be fitted with a theoretical curve where the substrate has index $\tilde{N}_s = 4.346 - 3.744$ j and the film has index $\tilde{N}_f = 2.392$. Using this, we can determine the thickness at each point and Figure C6-9 shows the thickness plotted versus the potential. Again, we obtain a linear plot indicating that the electric field in the film is independent of the thickness.

In the second paper,[8] Clayton and DeSmet deal with the fact that in Figure C6-8, the second loop does not exactly overlap the first loop. A closer inspection of this figure shows that the squares on the left-hand part of the figure are inside the first loop and the squares on the right side of the loop are on the outside part of the first loop.

We have seen several reasons why the second loop does not coincide with the first loop. In the first example, zirconium, we observed

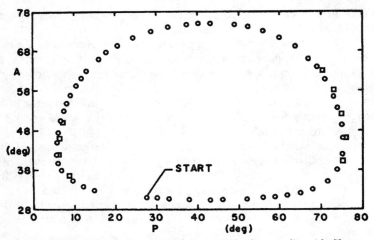

Figure C6-8. Ellipsometer null settings for the growth of an anodic oxide film on vanadium at a current density of 275 μA/cm². Open circles, first loop; open squares, second loop. (After Clayton[7])

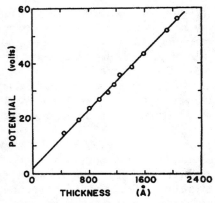

Figure C6-9. Potential vs. thickness for the anodic oxidation of vanadium at a current density of 275 μA/cm². Thicknesses were calculated from the data in Figure C6-8. (After Clayton[7])

an outward spiral, indicating a nonzero value of k. In the second example, titanium, we observed the trajectory to move inward for increasing thicknesses and this was attributed to a changing index. In the present case, the deviation is neither inward or outward; instead, the values of P are offset to a sightly higher value for each successive loop. We have not discussed a model that will explain this particular behavior.

Up to now, we have assumed that the index of refraction was isotropic. Let us consider the effect of an anisotropic index. The effect of uniaxial anisotropy with an optic axis parallel to the direction of film growth (perpendicular to the surface) is to displace successive loops parallel to the P axis by an amount that depends on the size of the anisotropy and in a direction that depends on whether the anisotropy is uniaxial positive or uniaxial negative.[8] For the configuration used by Clayton and DeSmet, a uniaxial negative film ("extraordinary" index, $\tilde{N}_e$ < "ordinary" index, $\tilde{N}_o$) will cause successive loops to be displaced toward higher values of P.

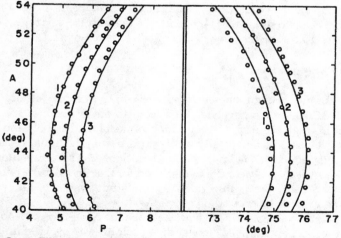

Figure C6-10. Ellipsometer null settings in the regions of minimum and maximum P for the growth of an anodic film on vanadium at a current density of 394 $\mu A/cm^2$. Data points are represented by circles and the solid lines represent the calculated behavior for a uniaxial film with an optic axis normal to the surface of the film for which $\tilde{N}_o$ = 2.349 and $\tilde{N}_e$ = 2.328. The refractive index of the substrate is assumed to be $\tilde{N}_s$ = 3.838 - 3.560 j. (After Clayton[8])

This displacement effect will have the most pronounced effect in regions where P is either a maximum or minimum. Figure C6-10 shows these regions of the P/A trajectory for anodic growth of a film using a constant current density of 394 $\mu A/cm^2$. Note that the middle part of the trajectory has been taken out of the figure. The numbers in the figures indicate the first, second, and third loops respectively. The offset toward higher values of P is more evident in this figure than in Figure C6-8. The solid lines represent calculations using a substrate index $\tilde{N}_s = 3.838 - 3.56$ j, an ordinary refractive index $\tilde{N}_o = 2.349$, and an extraordinary refractive index $\tilde{N}_e = 2.328$, giving a difference of $\Delta\tilde{N} = 0.021$.

This type of anisotropy has been observed with other materials as well. Tantalum[47] and niobium[48] each exhibit negative anisotropy with the amount of anisotropy depending on the electric field. For current densities in the range of 300 to 400 $\mu A/cm^2$ and for fields in the 4 to 6×10^6 V/cm, for tantalum, $\Delta\tilde{N} = 0.0039$ and for niobium, $\Delta\tilde{N} = 0.0056$. Clayton and DeSmet[8] find that for similar current density and lower field, 2.7×10^6 V/cm, the difference is significantly greater, i.e., $\Delta\tilde{N} = 0.021$. Note that these observations are made as the film grows, i.e., with the electric field present. For tantalum and niobium, it is observed that when the field is removed, the anisotropy disappears. Clayton and DeSmet go on to show[8] that for vanadium, the anisotropy is reduced, but does not go completely away when the field is removed.

C6.6　Deposition of Oxides: 1. Lead

In film growth, one of the species comes from the substrate and the other from the electrolyte. This requires one of the film species to travel through the film. In deposition of metal oxides, both the metal and oxygen atoms must arrive at the surface from the liquid phase and be converted to the solid phase. For deposition, the substrate functions only as an electrically conductive support for the film. For the two cases discussed below, the substrate was a platinum single crystal cylinder with an optical flat ground on one side.

Ord, Huang, and DeSmet[17] describe the deposition of lead dioxide films. As mentioned above, the substrate was a platinum single crystal and the deposition solution was a buffered lead acetate solution (1M in lead acetate, sodium acetate, acetic acid). The refractive index of the electrolyte is 1.3752. These conditions are reported[49] to be favorable for the deposition of alpha lead dioxide. The counter-electrode was platinum and the potential was measured with a mercury-mercurous sulfate reference electrode. A current density of 694 μA/cm^2 was used for this deposition.

The upper part of Figure C6-11 shows a plot of some of the P/A points. In this case, one-third of the points measured are displayed and they are taken at equal charge transfer increments. The trajectory starts at point A, which is representative of film-free platinum. The trajectory moves in a clockwise manner gradually spiraling outward. This type of curve is characteristic of a material that is slightly absorbing (k > 0).

In determining the values of n, k, and the thickness for the various points, it was assumed that the values of n and k should be the same for all thicknesses and that the relationship between charge/cm^2 transferred and thickness should be linear. The procedure for selecting the appropriate values of n and k is to assume a value for n and determine the resulting values of k and the thickness for the various points. The

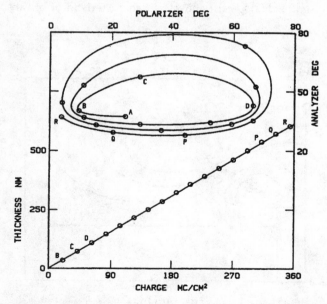

Figure C6-11. Galvanostatic deposition of PbO$_2$ onto a platinum substrate at 694 μA/cm^2. The circles identify every third point from the set used in the optical analysis, and the calculated thicknesses are for a film with Ñ = 2.03 - 0.045 j. (After Ord[17])

procedure was repeated for various values of n. The value of n that gives the lowest standard deviation for k and the smallest deviation from a linear charge/thickness is considered the best fit. The solid line in Figure C6-11 is the calculated trajectory for $\tilde{N}_f = 2.03 - 0.045\ j$. Using these values, the thicknesses were determined and the plot of charge/cm^2 versus thickness is shown in the lower part of the figure.

C6.7   Deposition of Oxides: 2. Manganese

As a second example of deposition of oxides, we use a work of Ord and Huang,[16] presented a little earlier in 1985 than the previous example. The same substrate was used. The films were deposited from an air saturated   solution of   0.1M MnSO$_4$ + 0.017M H$_2$SO$_4$. Figure C6-12 shows results from a deposition experiment at a current density of 44 µA/cm^2. The potential, shown in the lower part of the figure, remained constant. The P/A trajectory is shown in the upper part of the figure.   Again, this spiral behavior is characteristic of an absorbing

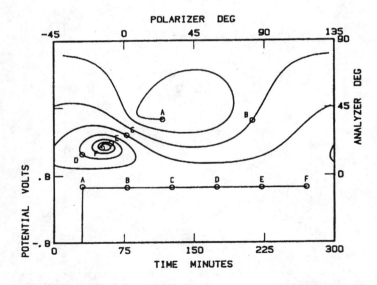

Figure C6-12. Galvanostatic deposition of manganese dioxide onto a platinum substrate at 44 µA/cm^2. (After Ord[16])

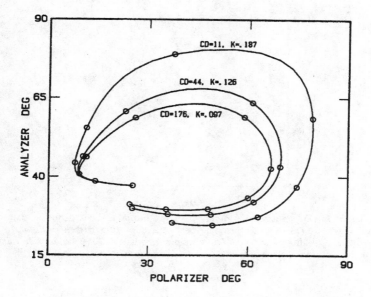

Figure C6-13. First-cycle optical data for galvanostatic deposition of manganese dioxide onto a platinum substrate at current densities from 11 to 704 $\mu A/cm^2$. The optical analysis uses a common value of 1.9 for n and finds the value of k that gives the best fit to the data. The cycle at 704 $\mu A/cm^2$ (with k = 0.096) overlaps the cycle at 176 $\mu A/cm^2$ and is not shown. (After Ord[16])

material. In this case, the film thickness becomes large enough that the spiral moves toward the bulk film point, i.e., the point characteristic of a very thick substrate made of the film material.

Deposition experiments at other current densities give optical results which are similar, in general, but different in specifics. They all start at the same point (characteristic of the Pt substrate) and spiral in the same general direction but they differ in the number of inner and outer loops and in the terminal point (the asymptote of the inner spiral). Figure C6-13 shows the first loop for three different current densities (labeled "CD"), ranging from 11 to 176 $\mu A/cm^2$. The curve for current density 704 $\mu A/cm^2$ (not shown) overlaps the 176 $\mu A/cm^2$ curve.

The same methods described in the previous example were used to determine n, k, and the thickness for the various points. The results

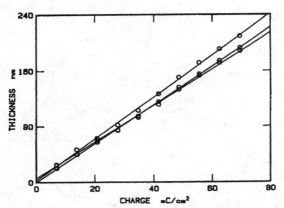

Figure C6-14. Dependence of film thickness on accumulated charge for the data from Figure C6-13. The slope is steepest at a deposition current density of 176 μA/cm², shallowest at 11 μA/cm², and intermediate at 704 μA/cm². The data at 44 μA/cm² overlap the data at 176 μA/cm² and are not shown. (After Ord[16])

were that the value of n, 1.9, was independent of the current density but the values of k (labeled "K") decrease as the current density increases. For the current densities used, k ranged form 0.097 to 0.187. The theoretical curve shown in Figure C6-12 corresponds to $\tilde{N}_f = 1.9 - 0.126\,j$. Using these index values, the thicknesses were calculated and plotted versus charge accumulation in Figure C6-14 for three of the four conditions discussed above.

C6.8   References

1. J. L. Ord and B. L. Wills, *Appl. Opt.*, 6 1673 (1967).
2. J. L. Ord, *Surface Sci.*, **16**, 155 (1969).
3. M. A. Hopper and D. J. DeSmet, *Bull. Am. Phys. Soc.*, **15**, 1326 (1970).
4. J. L. Ord, M. A. Hopper, and W. P. Wang, *J. Electrochem. Soc.*, **119**, 439 (1972).
5. M. A. Hopper and J. L. Ord, *J. Electrochem. Soc.*, **120**, 183 (1973).
6. J. L. Ord, *Surface Sci.*, **56**, 423 (1976).
7. J. C. Clayton and D. J. DeSmet, *J. Electrochem. Soc.*, **123**, 174 (1976).
8. J. C. Clayton and D. J. DeSmet, *J. Electrochem. Soc.*, **123**, 1886 (1976).
9. M. A. Hopper, J. A. Wright, and D. J. DeSmet, *J. Electrochem. Soc.*, **124**, 44 (1977).
10. J. L. Ord, J. C. Clayton, and D. J. DeSmet, *J. Electrochem. Soc.*, **124**, 1714 (1977).
11. J. L. Ord, J. C. Clayton, and K. Brudzewski, *J. Electrochem. Soc.*, **125**, 908 (1978).
12. J. L. Ord and E. M. Lushiku, *J. Electrochem. Soc.*, **126**, 1374 (1979).

13. J. L. Ord, *J. Electrochem. Soc.*, **129**, 335 (1982).
14. J. L. Ord, *J. Electrochem. Soc.*, **129**, 767 (1982).
15. Z. Q. Huang and J. L. Ord, *J. Electrochem. Soc.*, **132**, 24 (1985).
16. J. L. Ord and Z. Q. Huang, *J. Electrochem. Soc.*, **132**, 1183 (1985).
17. J. L. Ord, Z. Q. Huang, and D. J. DeSmet, *J. Electrochem. Soc.*, **132**, 2076 (1985).
18. J. L. Ord, D. J. DeSmet, and Z. Q. Huang, *J. Electrochem. Soc.*, **134**, 826 (1987).
19. D. J. DeSmet and J. L. Ord, *J. Electrochem. Soc.*, **134**, 1734 (1987).
20. D. J. Beckstead, G. M. Pepin, and J. L. Ord, *J. Electrochem. Soc.*, **136**, 362 (1989).
21. D. J. Beckstead, D. J. DeSmet, and J. L. Ord, *J. Electrochem. Soc.*, **136**, 1927 (1989).
22. J. L. Ord, D. J. DeSmet, and D. J. Beckstead, *J. Electrochem. Soc.*, **136**, 2178 (1989).
23. D. J. DeSmet and J. L. Ord, *J. Electrochem. Soc.*, **136**, 2841 (1989).
24. G. R. J. Robertson, J. L. Ord, D. J. DeSmet, and M. A. Hopper, *J. Electrochem. Soc.*, **136**, 3380 (1989).
25. J. L. Ord, S. D. Bishop, and D. J. DeSmet, *J. Electrochem. Soc.*, **138**, 208 (1991).
26. C. Lukac, J. B. Lumsden, S. Smialowska, and R. W. Staehle, *J. Electrochem. Soc.*, **122**, 1571 (1975).
27. Z. Szklarska-Smialowska, and W. Kozlowski, *J. Electrochem. Soc.*, **131**, 499 (1984).
28. W. Kozlowski and A. Szklarska-Smialowska, *J. Electrochem. Soc.*, **131**, 723 (1984).
29. T. Zakroczymski, Chwei-Jer Fan, and Z. Szklarska-Smialowska, *J. Electrochem. Soc.*, **132**, 2862 (1985).
30. K. Sugimoto and S. Matsuda, *J. Electrochem. Soc.*, **130**, 2232 (1983).
31. K. Sugimoto, S. Matsuda, Y. Ogiwara, and K. Kitamura, *J. Electrochem. Soc.*, **132**, 1791 (1985).
32. X. Shan, D. Ren, P. Scholl, and G. Prentice, *J. Electrochem. Soc.*, **136**, 3594 (1989).
33. J.-G. Hwu and M.-J. Jeng, *J. Electrochem. Soc.*, **135**, 2808 (1988).
34. T. Smith, P. Smith, and F. Mansfeld, *J. Electrochem. Soc.*, **126**, 799 (1979).
35. S. Silverman, G. Cragnolino, and D. D. Macdonald, *J. Electrochem. Soc.*, **129**, 2419 (1982).
36. T. Ohtsuka, M. Masuda, and N. Sato, *J. Electrochem. Soc.*, **132**, 787 (1985).
37. R. Greef and C. F. W. Norman, *J. Electrochem. Soc.*, **132**, 2362 (1985).
38. C. K. Dyer and J. S. L. Leach, *J. Electrochem. Soc.*, **125**, 23 (1978).
39. P. W. T. Lu and S. Srinivasan, *J. Electrochem. Soc.*, **125**, 1416 (1978).
40. A. Moritani, H. Kubo, and J. Nakai, *J. Electrochem. Soc.*, **126**, 1191 (1979).
41. F. M. Cain, "Zirconium and Its Alloys," p. 176, American Society for Metals, Materials Park OH (1953).
42. J. Pelleg, *J. Less-Common Met.*, **35**, 299 (1974).
43. T. Smith and F. Mansfeld, *J. Electrochem. Soc.*, **119**, 663 (1972).
44. J. H. Richardson, "Optical Microscopy for the Materials Sciences," pp. 406-407, Marcel Dekker, New York (1971).
45. R. G. Keil and R. E. Salomon, *J. Electrochem. Soc.*, **112**, 643 (1965).
46. R. G. Keil and R. E. Salomon, *J. Electrochem. Soc.*, **115**, 628 (1968).
47. W. D. Cornish and L. Young, *Proc. R.Soc. London A*, **335**, 39 (1973).
48. K. K. Yee and L. Young, *Appl. Opt.*, **14**, 1316 (1975).
49. H. H. Johnson and A. M. Willner, *Appl. Mater. Res.*, **4**, 35 (1965); E. M. Hackett, J. R. Scully, and P. J. Moran, Paper 136 presented at the Electrochemical Society Meeting, Washington, DC, Oct. 9-14, 1983; J. J. Miksis, and J. Newman, *J. Electrochem. Soc.*, **123**, 1030 (1976).
50. J. L. Ord and D. J. DeSmet, *J. Electrochem. Soc.*, **139**, 359, 728 (1992).

Case 7:
Amorphous Hydrogenated Carbon Films

C7.1   General

Insulator films are used extensively on semiconductors, particularly in integrated circuits. SiO_2, Si_3N_4, and polyimide are used extensively in silicon based structures. For the Group III-V based structures, however, there has been an effort[1-3] to develop an amorphous hydrogenated carbon (a-C:H) insulator film.   Attractive properties are[2] that:

1. It is easily prepared as a homogeneous film.
2. It is very hard mechanically and chemically.
3. It has a high breakdown voltage and resistivity.
4. It has a relatively low interface density of states on Si and InP.
5. It has a variable energy gap.
6. It can be used in metal-insulator-metal structures.

Many of these properties are attributed to a large number of tetrahedral (sp^3) bonds and thus it is referred to as "diamond-like."[4,5] For weakly absorbing a-C:H films, a strong correlation has been noted between diamond-like physical and optical properties.[5,6] The film is normally deposited by a plasma enhanced chemical vapor deposition process using a hydrocarbon gas.

C7.2   Mechanism of Film Formation

General

Kersten and Kroesen[1] investigated the mechanism of film formation. The two mechanisms in question were direct incorporation of

160

particles from the gas phase (flow-in model) and deposition from an adsorbed layer (adsorbed-layer model).[7] The key to distinguishing between these two models is the temperature dependence of the deposition rate.[1] A positive temperature slope indicates that the particles are directly incorporated in the film upon chemisorption. A negative slope indicates that the incorporation is intermediated by an adsorbed layer.

Experimental

The deposition reactor is shown in Figure C7-1. A thermal plasma generated with a cascaded arc is employed to dissociate and ionize molecular gases. The complete arc system is fixed to an anode supporting flange, which is attached to a vacuum system. Methane is dissociated in the arc and the products expand into the vacuum system and are deposited on the surface. When the high velocity ions and neutrals collide with the surface, these particles transfer their energy into heat and elevate the substrate temperature.

The substrates were small glass or steel plates coated with a thin layer of gold. The standard process parameters were:

arc current, 50 A,

arc voltage, 85 V,

arc pressure 4.7×10^4 Pa,

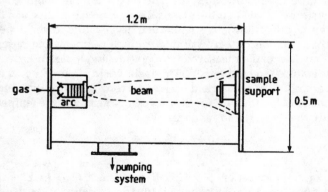

Figure C7-1. Outline of the reactor used in the present experiments. The gas is fed through a cascaded arc and emanates in the vacuum system where an intense plasma beam is created. The thus-created reactive particles are deposited on the substrate, which is mounted on the sample support. (After Kersten[1])

argon flow, 100 std. cm^2/s,
methane flow, 1.86 std. cm^2/s,
background pressure, 10^{-3} Pa
process pressure 20 to 250 Pa.

Although water cooling of the substrate was used on occasion, there was no attempt to hold the substrate temperature at a fixed value. The temperature was recorded as a function of process duration, and was measured with thermocouples connected to the substrate but shielded from the plasma. The temperature increase during the process was usually no more than 30°C. Because of these variations, deposition occurred at many different substrate temperatures. *In situ* ellipsometry measurements were made simultaneously with temperature data acquisition. The ellipsometer used a wavelength of 6328 Å and presumably an angle of incidence of 70°.

Results

A typical Δ/Ψ plot is shown in Figure C7-2. The solid line is the calculated fit to the data. The gradual spiral outward implies a slightly absorbing film. The authors do not give index of refraction data, but the starting point, $\Delta/\Psi = 108.7°/43.8°$, taken from the plot matches that of gold reasonably well. The curve can be approximated by using an index of $\tilde{N} = 1.8 - 0.03j$.

Figure C7-3 shows film growth versus time for both glass and steel substrates. Note that the glass substrate, which will be at a higher temperature due to lower heat conductivity, has the lower deposition rate. The instantaneous growth rate from ellipsometry measurements can be matched to the measured temperature and a plot of growth rate vs. temperature can be obtained. This is shown in Figure C7-4 where the data are plotted as an Arrhenius plot. From this plot we obtain an activation energy of 0.58 eV for the process. Note that as the temperature increases, the deposition rate decreases.

Summary

For the deposition of a-C:H films from a CH_4-Ar plasma, we obtain a negative slope of the temperature dependence of the deposition rate. This indicates that adsorption-desorption equilibrium is the rate limiting step. We can conclude that the incorporation of reactive particles into the growing film may be intermediated by an adsorbed layer.[1]

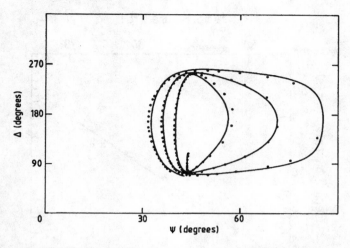

Figure C7-2. An example of Δ/Ψ curves as measured (points) with *in situ* ellipsometry during the deposition of amorphous carbon films on gold (on glass). The simulation used to determine the film thickness is also shown (full curve). (After Kersten[1])

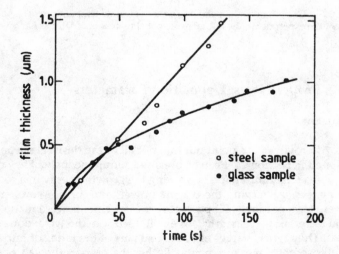

Figure C7-3. Film thickness vs. time for two sample materials. (After Kersten[1])

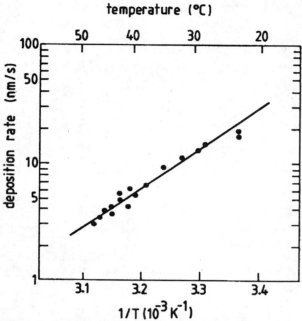

Figure C7-4. Growth rate in dependence on the substrate temperature. The points represent the experiments, the line an exponential regression. (After Kersten[1])

C7.3   Properties vs. Deposition Parameters

Experimental

     Alterovitz *et al.*[2] point out that not all of the desirable properties mentioned in Section C7.1 can be obtained simultaneously. They report a study of a-C:H films made by varying the growth energy in a continuous way, i.e., by varying the plasma power, and/or pressure, and/or substrate temperature. Ellipsometric measurements were made using several different wavelengths. We will focus on the work done using 6328 Å. They report values of n, the real part of the index of refraction, from ellipsometry measurements. Rather than report the values of the extinction coefficients from the ellipsometry measurements, they report absorption coefficient data from UV-visible absorption spectroscopy and their emphasis is in the UV region. We will not discus the absorption coefficient values other than to indicate that the resulting extinction coefficient values are not drastically different from our estimate in the previous discussion.

The a-C:H films were prepared by 30 kHz plasma deposition using methane gas. The substrates were positioned on the grounded (anode) electrode, while a negative dc voltage existed on the other parallel and equal-size top electrode. Substrates used included InP for ellipsometry and Si for deposition temperature dependency.

Results

Films were deposited using six values of the deposition power at a methane flow rate of 70 sccm, corresponding to a pressure of 315 mTorr. Figure C7-5 shows the results. Clearly, as the power is increased, the refractive index n increases. There is no information about how the substrate temperature is controlled, but Figure C7-6 shows the effect. Alterovitz *et al.*[2] feel that the last point is anomalous and that the

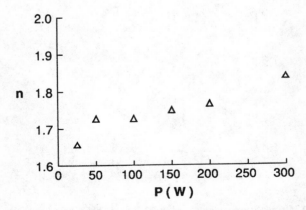

Figure C7-5. Refractive index n vs. deposition power P for a-C:H, CH_4 plasma deposited film on InP at 315 mTorr. (Data taken from Alterovitz[2])

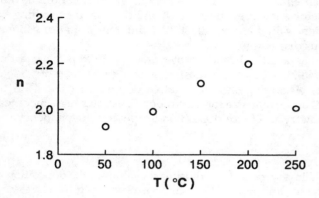

Figure C7-6. Refractive index n vs. substrate temperature for a-C:H, CH$_4$ plasma deposited film on Si at 100 W, 245 mTorr. (Data taken from Alterovitz[2])

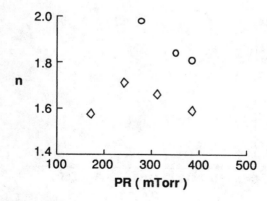

Figure C7-7. Refractive index n vs. pressure PR for a-C:H, CH$_4$ plasma deposited film on InP at two deposition powers P. The circles represent 240 W power and the diamonds represent 50 W power. (Data taken from Alterovitz[2])

deposition process is somewhat changed above 200°C. With this in mind, then, we see that increasing substrate temperature increases the refractive index.

Figure C7-7 shows the effect of pressure. For the higher pressures, increased pressures causes decreased index values. Note that the higher the pressure, the lower the average energy per particle, so higher pressure implies lower energy.

Summarizing, higher power, higher substrate temperature and lower pressure (at least, above 250 mTorr) gives higher values for the index of refraction. Collins[3] indicates that an index of refraction is an indicator of the polymer-like or diamond-like nature of the material, with higher n implying more diamond-like character.

C7.4   References

1. H. Kersten and G. M. W. Kroesen, *J. Vac. Sci. Technol. A*, **8**, 38 (1990).
2. S. A. Alterovitz, J. D. Warner, D. C. Liu, and J. J. Pouch, *J. Electrochem. Soc.*, **133**, 2339 (1986).
3. R. W. Collins, *J. Vac. Sci. Technol. A*, **7**, 1378 (1989).
4. J. Robertson, *Adv. Phys.*, **35**, 317 (1986).
5. J. Angus, P. Koidl, and S. Domitz, in "Plasma Deposited Thin Films," edited by J. Mort and F. Jensen, CRC, Boca Raton, FL (1986), p. 89.
6. A. Bubenzer, B. Dischler, G. Brandt, and P. Koidl, *J. Appl. Phys.*, **54**, 4590 (1983); B. Dischler, R. E. Sah, P. Koidl, W. Fluhr, and A. Wokaun, in "Proceedings of the 7th International Symposium on Plasma Chemistry," edited by C. J. Timmermans, Technical University of Eindhoven, The Netherlands (1985), p. 45; F. W. Smith, *J. Appl. Phys.*, **55**, 764 (1984); N. Savvides, *J. Appl. Phys.*, **59**, 4133 (1986).
7. H. Deutsch and M. Schmidt, *Beitr. Plasmaphys.*, **21**, 279 (1981).

Case 8:
Fluoropolymer Films on Silicon from Reactive Ion Etching

C8.1   General

Reactive ion etching (RIE) is an important method of transferring a pattern from photoresist to the material in an evolving integrated circuit. CF_4, a fluorocarbon, is often used with differing amounts of hydrogen or oxygen. By increasing the amount of hydrogen, the etch process can be made more selective to SiO_2 over Si.[1] It has been determined[2] that a fluorocarbon film is deposited onto the surface, which prevents the fluorine from attacking the Si. The fluorocarbon film does not form on SiO_2 presumably because the oxygen freed during etching reacts with the film precursor.[3] The fluoropolymer is therefore an intrinsic part of the selective etching process.[4] There have been several studies of this fluorocarbon polymer[3-6] and we shall discuss some of the results obtained using ellipsometry as a measurement technique.

C8.2   Film Formation and Properties

Refractive Index Studies

In 1986 Oehrlein et al.[4] published the first of a series of papers. In this work, the optical constants and film composition are investigated. Films were deposited under a variety of RIE conditions using CF_4 with several different fractions of hydrogen added. The first samples used were silicon <100> wafers, which had been chemically cleaned just prior to film formation. The plasma parameters were 0.18 W/cm^2, 40 sccm,

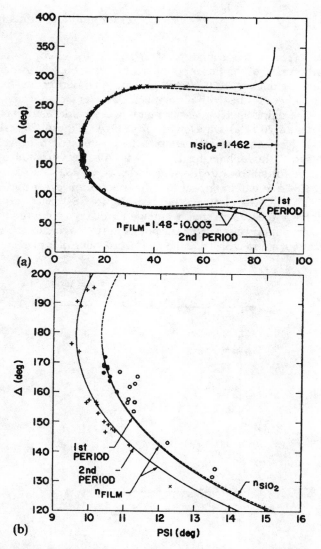

Figure C8-1. (a) Ellipsometry data for a number of fluorocarbon films deposited onto silicon under various reactive ion etching conditions: Solid circles, CF_4 -X %H_2; open circles, CF_4 - 40% H_2, x, +, CF_4-50%H_2. The calculated curve (solid line) for $\tilde{N}_f = 1.48 - 0.003j$ is shown. For comparison, a curve for SiO_2 is also shown (dashed line). (b) Expanded view of the Δ/Ψ trajectory for very thin films of film thicknesses close to the period thickness. The Δ/Ψ data of fluorocarbon films are not exactly periodic, which requires the complex film refractive index. (After Oehrlein[4])

and 25 mTorr using a mixture of CF_4 with 50% H_2 by volume. The deposition rate of the fluoropolymer film was about 60 Å per minute, and films as thick as 3000 Å were grown. As described in Chapter 4, a large number of films with different thicknesses were formed.

The ellipsometry measurements were made using an angle of incidence of 70° and wavelength of 6328 Å. Figure C8-1 shows the results. Note that several samples had values of Δ near 360° (the "sweetspot"), thus enhancing the analysis. Also drawn on the upper part is a curve for silicon dioxide with index Ñ = 1.462. At first glance at Figure C8-1a, it would appear that the trajectory is cyclic and hence the extinction coefficient k would be zero. Figure C8-1b is an expansion of the plot near the film free point, and we see that the points on the second period are slightly offset from the points near zero thickness. This is the trajectory of a film with a small but nonzero value of k. Using the index

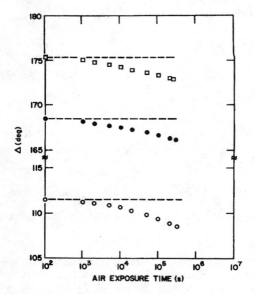

Figure C8-2. Change in ellipsometric variable Δ as a function of the air exposure time of reactively ion-etched silicon as a result of oxygen permeation of the fluorocarbon film and growth of an interfacial oxide layer. Squares, 100% CH_4; filled circles, CF_4-30%H_2; open circles, CF_4-40%H_2. (After Oehrlein[4])

of the silicon substrate as $\tilde{N}_s = 3.85 - 0.025j$, the solid line is the best fit, obtained by using a film index of $\tilde{N}_f = 1.48 - 0.003j$.

A better fit of these data can be obtained if the refractive index used for the substrate is changed. The physical reason for this is probably silicon substrate damage which occurred during the RIE process. Note that we would expect that, as we change the fraction of hydrogen in the RIE process, the composition of the film would change. Oehrlein et al.[4] conclude that the value of the index of the film is somewhat insensitive to composition.

Air exposure of the dry-etched samples causes changes in Δ because of oxygen permeation to the fluoropolymer/substrate interface followed by oxide growth which is enhanced due to the RIE damage. Some of the changes in Δ are shown in Figure C8-2. Using the previously determined index values and taking the oxide/polymer to be a single film, it is observed that the sample etched in 100% CF_4 changes from 2.5 Å immediately after RIE to 11 Å within 3.5 days. The sample etched with 40% H_2 changed from 306 Å to 330 Å. Because of this, the air exposure between deposition and measurement was kept constant.

Helium ion backscattering analysis was used to measure the ratio of carbon to fluorine for a large number of the films and the results are shown in Figure C8-3. Very thin films have a large carbon/fluorine ratio indicating that these films contain very little fluorine. The thicker films approach a 1:1 ratio, indicating much more fluorine. It is clear that the thicker films are not simply larger versions of the thinner films. With this in mind, one wonders why it is possible to fit one curve to the data for all of the various thicknesses. The reason is that the index is somewhat insensitive to these variations. If one were to assume that the index was 1.8 when it was actually 1.48, the error in thickness would be less than 15% for film thicknesses up to 100 Å.

Several samples with 500 Å oxide were etched in various H_2 fractions until the oxide was removed. The RIE was continued for a 1 min over-etch to form the fluoropolymer film. For a constant exposure time, the thickness of the fluoropolymer film increases with increasing H_2 concentration in the feed gas. Figure C8-4 shows the result of this experiment. In this figure, the thickness as determined with the index discussed above is plotted as a function of the areal density of both fluorine and carbon (determined by helium backscattering). Assuming a constant density of the residue films, we would expect to see a linear increase in film thickness for increasing areal densities, and this is observed.

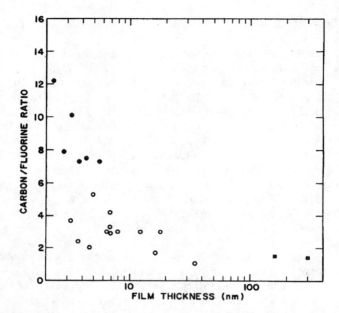

Figure C8-3. Carbon-to-fluorine ratio for fluorocarbon films deposited under widely differing RIE conditions vs. the ellipsometric film thickness: Closed circles, CF_4-X%H_2; open circles, CF_4-40%H_2; closed squares, CF_4-50%H_2. (After Oehrlein[4])

The thicknesses of some of the thicker films were measured with a stylus instrument. The stylus results were within 3% or less of the ellipsometry results.

C8.3   Interface Studies

In 1988, Jaso and Oehrlein[3] published a work where they extended the previous studies.[4] In this work, they used a different ellipsometer, which used an angle of incidence of 75° along with the usual wavelength of 6328 Å. Si <100> wafers covered with 1000 Å of thermal oxide were used as samples. For the RIE, the power was 200 W, the gas pressure was 25 mTorr, and the total gas flow was 20 sccm. The compositions of the gas were CF_4 with 10, 20, and 45% H_2 and pure CHF_3. Most of the wafers were etched for 11 min.

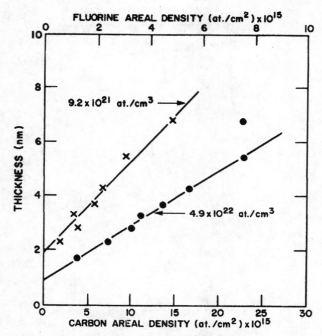

Figure C8-4. Ellipsometric film thickness vs. RBS carbon (closed circles) and fluorine
(x) areal densities for several plasma-exposed (due to overetching) silicon samples.
A CF_4-$X\%H_2$ gas mixture ($X \leq 40\%$) was used for RIE. The changing composition of
the films as a function of thickness is borne out by the different slopes of the two
straight lines. (After Oehrlein[4])

For the ellipsometry analysis, we note from the previous work
that for films less than 100 Å thick that only the value of Δ changes
significantly. Accordingly, in this work only Δ is plotted as a function of
time and processing conditions. When the film thickness decreases as a
function of time, we expect that Δ will increase as a function of time.

In Figure C8-5, we see how Δ changes as a function of time. The
initial rise represents the removal of the oxide layer. Δ approaches a
value of 141°, which is characteristic of the clean Si substrate for the
experimental conditions used in this work.[3] Subsequently, Δ drops
again due to the formation of the fluoropolymer film after the oxide has

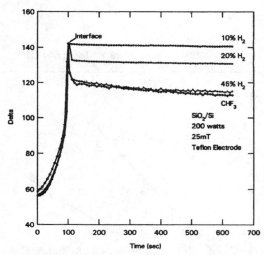

Figure C8-5. Δ measured as SiO_2/Si is reactive ion etched in CF_4/H_2 using various percentages of H_2 and CHF_3 vs. time. (After Jaso[3])

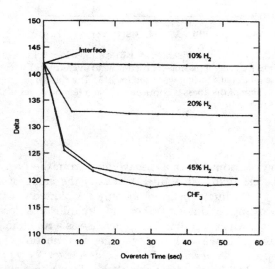

Figure C8-6. Δ as a function of overetching time measured from the encounter of the SiO_2/Si interface for CH_4/H_2 and CHF_3 plasmas. Fluorocarbon film deposition is indicated by a decrease in Δ. (After Jaso[3])

been removed. For 10% H_2, very little film forms. For increased percentages of H_2, however, a thicker film is formed. Note that steady-state is reached very rapidly. Figure C8-6 is an expanded view of the initial part of the curve immediately after etching through the oxide film. Within 15 s films formed with all of the plasma conditions have reached a near steady-state thickness. If a refractive index of 1.462 is assumed, the film thickness is 6 Å for the 20% H_2 condition and 17 Å for the other two conditions.

Note that in Figures C8-5 and C8-6, the plasma continued. The authors[3] investigated whether the state of the specimen surface differs under plasma-on conditions from plasma-off conditions. A CF_4-20%H_2 plasma was used and after removing the oxide layer, the plasma was alternatively turned off and on while keeping the gas flow and pressure unchanged. Figure C8-7 shows the results. Note that when the plasma was turned off after reaching steady-state conditions, additional film

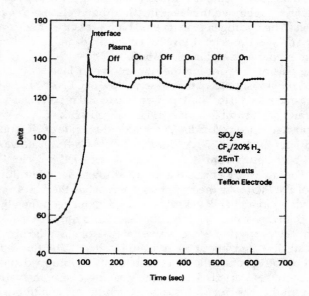

Figure C8-7. Δ measured as a function of time during reactive ion etching while turning the plasma off and on, demonstrating fluorocarbon film growth with the plasma off and the etching back of the film with the plasma on. (After Jaso[3])

was formed. When the plasma was turned on again, the thickness was reduced to the previous steady-state level. The plasma-on value of Δ was 130.7° whereas the plasma-off value was 126.5°. With the index assumption listed previously, this represents film thicknesses of 6 and 9 Å, respectively. Clearly, the steady-state layer represents an equilibrium of film formation and film removal.

The authors made one additional observation. They surrounded the wafer that had 1000 Å oxide with wafers with 1500 Å oxide. This inhibited the growth of the fluoropolymer film until the surrounding wafers were oxide-free.

C8.4   Removal with a Hydrogen Plasma

In a subsequent work[5] published in 1991, Simko et al. conducted a study of removing the fluoropolymer film with a hydrogen plasma. There have been various studies of removal of the fluoropolymer using oxygen.[7] All of these methods result in reoxidizing the Si surface. Simko et al.[5] attempts to remove the layer in a hydrogen plasma to avoid the reoxidation. The hydrogen plasma uses high pressures, high hydrogen pressures and gas flows, and low RF power in order to minimize ion bombardment damage and implantation.

The plasma for forming the fluoropolymer layer is the same as mentioned previously except that 40 sccm flow of CF_4-40% H_2 was used. The H_2 plasma used 70 to 100 W power, 370 to 400 mTorr, and 150 sccm H_2 flow rate. The samples were <111> silicon with 6500 Å chemical vapor deposited silicon dioxide. The average etch rate of SiO_2 under the RIE conditions used was 270 Å/min. In Figure C8-8 we show one part of a three-part figure from Ref. 5. This is the Δ/Ψ trajectory for the process. The trajectory enters the figure at the lower right during the removal of the SiO_2. As the trajectory approaches the film-free condition, it deviates to higher Ψ values due to RIE damage. When the oxide is removed, the over-etch and the fluoropolymer film deposition begin and the trajectory drops to lower values of Δ. Note that the film growth slows drastically after about 30 Å are deposited. The fluoropolymer reaches about 50 Å before the film-formation RIE is stopped and the H_2 plasma cleaning is started. During the cleaning phase, the values of Δ increase until the trajectory approaches the film-free condition.

Figure C8-9 shows a second part of the three-part figure in Ref. 5. This is simply the values of Δ from Figure C8-8 plotted as a function of time. The nearest that the trajectory approaches the film-free condition is about 10 Å. The cleanest point is reached in 20 min and corresponds to a cleaning rate of about 2 Å/min.

C8.5   Other Formation and Removal Studies

Flowers et al.[6] published a work in 1990 where they were able to deposit the fluoropolymer on SiO_2. The RIE gas was CHF_3 and they found that with a pressure of 200 mTorr and a power setting of 50 W, deposition occurred, whereas when the power setting was raised to 200 W, the fluoropolymer was removed. After the fluoropolymer was removed, the plasma then began to remove the underlying oxide.

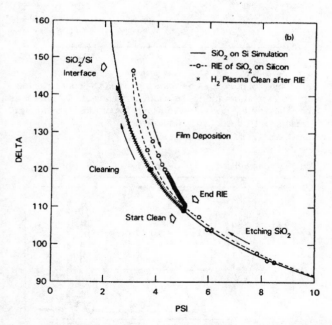

Figure C8-8. Δ/Ψ plot of RIE of SiO_2 from silicon (closed circles past encounter of the SiO_2/Si interface to a 5 min over-etch point and the subsequent H_2 plasma clean (x). Solid line is a calculated Δ/Ψ trajectory for SiO_2 on Si. Plasma clean parameters were 80 W RF, 390 mTorr, and 150 sccm H_2 flow. (After Simko[5])

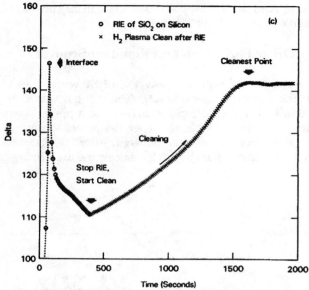

Figure C8-9. A Δ-time plot of Figure C8-8 showing the SiO₂/Si interface, the deposition of fluorocarbon film, and the subsequent clean by the H₂ plasma as a function of time. The cleaning rate is 2 Å/min and 10 Å of film is left on the Si surface when cleaning is stopped. (After Simko[5])

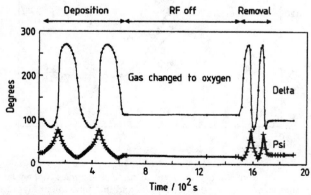

Figure C8-10. Ellipsometric readings (points) during polymer deposition followed by polymer removal by plasma etching in 300 mTorr O_2 at 150 W. (After Flowers[6])

They also used an oxygen plasma to remove the fluoropolymer. Their O_2 plasma conditions were pressure, 300 mTorr; RF power, 150 W; gas flow rate, 80 sccm. Figure C8-10 shows first the deposition of the fluoropolymer on SiO_2 and then the removal of the fluoropolymer with the oxygen plasma. In this case, when the fluoropolymer is removed, the oxygen plasma does not affect the underlying oxide.

C8.6   References

1. J. W. Coburn, "Plasma Etching and Reactive Ion Etching," American Vacuum Society Monograph Series, American Vacuum Society, New York (1982); R. A. Heinecke, *Solid State Electron.*, **18**, 1146 (1975).
2. G. S. Oerhlein and H. L. Williams, *J. Appl. Phys.*, **62**, 662 (1987); G. S. Oerhlein, S. W. Robey, and M. A. Jaso, *Mater. Res. Soc. Symp. Proc.*, **98**, 229 (1987).
3. M. A. Jaso and G. S. Oehrlein, *J. Vac. Sci. Technol. A*, **6**, 1397 (1988).
4. G. S. Oehrlein, I. Reimanis, and Young H. Lee, *Thin Solid Films*, **143**, 269 (1986).
5. J. P. Simko, G. S. Oehrlein, and T. M. Mayer, *J. Electrochem. Soc.*, **138**, 277 (1991).
6. M. C. Flowers, R. Greef, C. M. K. Starbuck, P. Southworth, and D. J. Thomas, *Vacuum*, **40**, 483 (1990).
7. G. S. Oehrlein, J. G. Clabes, and P. Spirito, *J. Electrochem. Soc.*, **133**, 1002 (1986); S. J. Fonash, X. C. Mu, S. Chakravarti, and L. C. Rathbun, *J. Electrochem. Soc.*, **135**, 1037 (1988); D. J. Vitkavage and T. M. Mayer, *J. Vac. Sci. Technol. B*, **4**, 1283 (1986); G. S. Oehrlein, G. J. Scilla, and S. J. Jeng, *Appl. Phys. Lett.*, **52**, 907 (1988).

Case 9:
Various Films on InP

C9.1 General

InP is finding applications in optoelectronics,[1] photoelectrochemistry,[2] and microwave devices.[3] Several works[3-7] have used ellipsometry to study the surface of InP and films on this material. We shall start with an investigation of the optical properties of the surface[3] itself. Then we will consider the thermal oxide that grows on InP.[4] This will be followed by a discussion of polymethyl methacrylate (PMMA) films on InP.[5]

C9.2 The InP Surface Optical Properties

Liu *et al.*[3] conducted a study of the surface optical properties of InP. Commercially available polished N-type Sn-doped <100> oriented

TABLE C9-1. Summary of cleaning procedures for InP surface (Information from Liu[3])

Group	Cleaning Procedure
1	Degreasing: 5 min in hot tetrachloroethylene, 10 min in 1:1 acetone:methanol, 10 min in acetone, N_2 dry.
2	Degreasing, 10 s concentrated HF dip, DI water rinse, N_2 dry.
3	Degreasing, 1% Br_2 in methanol chemo-mechanical polish, methanol rinse, N_2 dry.
4	Degreasing, 10 s HF dip, DI water rinse, 1% Br_2 in methanol chemo-mechanical polish, methanol rinse, N_2 dry.

InP wafers were used in the surface cleaning experiments. The surface measured by ellipsometry depends on the surface preparation method. Four cleaning procedures were considered and they are listed in Table C9-1. The Group 1 procedure consisted of a 5 min boiling in tetrachloroethylene followed by rinsing in a 1:1 mixture of acetone and methanol followed by 10 more minutes in acetone, followed by a N_2 dry. The Group 2 procedure consisted of the first part of the degreasing procedure listed above, followed by 10 s in concentrated HF, followed by a DI rinse and N_2 dry. The Group 3 procedure consisted of degreasing followed by a chemo-mechanical polish in 1% Br_2 in methanol, followed by a methanol rinse and N_2 dry. The Group 4 procedure was a combination of all of the above.

Immediately after the cleaning procedure, ellipsometry measurements were made on two different spots on the sample surface. A manual ellipsometer was used with angle of incidence of 70° and a wavelength of 6328 Å. The measured Δ and Ψ values are dependent on the surface treatment. The values measured are listed in Table C9-2. The values of Δ differed by as much as 3.6° whereas Ψ differed by only 0.25°. For a film with index in the range of 1.46 to 2.0, we can calculate using the program in Appendix A that this difference represents about 10 to 13 Å.

The significance of these differences is that it is not clear how much of a native oxide film covers the InP. If one were considering *in situ* experiments, this would be a large uncertainty. Compared with the certainty of the values of the index of silicon, this is also a large uncertainty. From the point of view of *ex situ* studies of film growth or

TABLE C9-2. Ψ and Δ values for InP surfaces after receiving various surface treatments as described in Table I. (Data from Liu[3])

Sample preparation	Ψ	Δ
No Clean	8.55°	155.81°
Group 1	8.50	155.25
Group 2	8.34	157.51
Group 3	8.31	158.94
Group 4	8.31	157.70

deposition which are several hundred angstroms thick, this difference is probably within spot-to-spot variations.

The authors conclude that the best value for the index of InP is $\tilde{N}_s = 3.521 - 0.300\,j$. This is consistent with a native oxide film for all of the above treatments which is 10 to 20 Å thick.

C9.3   The Thermal Oxide on InP

In a second work,[4] Liu *et al.*, studied the oxidation of InP. N-type undoped <100> InP wafers were given a treatment similar to that described for Group 2 above. They were oxidized as follows: 15 min preoxidation annealing at 440°C in N_2, oxidized in 1 atm O_2 at 440°C for 4 h, and annealed at 440°C for 15 min in N_2. This generated a film that was later determined to be between 550 and 600 Å thick. This film was then sequentially etched and measured with ellipsometry. The etchant

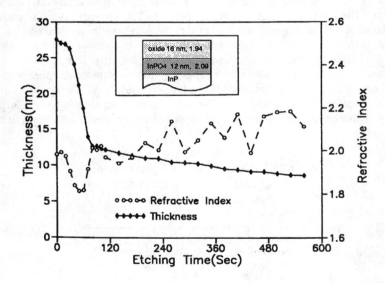

Figure C9-1. The thicknesses and refractive indices for the thermal InP oxide, etched by the dilute HF solution. The insert shows the two-layer structure for the oxide. (After Liu[4])

used was a 1:500 HF(49%):H_2O solution. The samples were dipped in the etchant for a certain time, rinsed in DI water, and blown dry with N_2. A slot on the sample stage ensured that the same surface spot could be probed for successive measurements.

The chemical composition of the oxide has been found[8,9] to be both complicated and process-dependent. It is generally accepted[10,11] that the oxide grown at temperatures below 600°C is composed of two layers of differing chemical composition: an In-rich outer layer of a mixture of In_2O_3 and $InPO_4$, and an inner layer of predominantly $InPO_4$.

Even knowing that this is a two-film situation, the ellipsometry data were analyzed first by assuming a single transparent film on a substrate with index $\tilde{N}_s = 3.521 - 0.300$ j. The thickness and refractive index values obtained are shown in Figure C9-1. It is clear from the figure that the slope of the thickness curve changes dramatically at around 90 s. Using the single-film model, the etch rate before the break point at 90 s is about 170 Å/min and the etch rate after the break point is about 5 Å/min.

It is clear that the breakpoint in the curve represents the interface between two different materials. Note that although the single film model is clearly not correct before 90 s of etch time, after this initial material is removed, a single-film model is appropriate for the analysis of the remainder of the film. The average index of refraction for the inner single film is $\tilde{N}_{f1} = 2.09$. One can then go back and reanalyze the data for the lower etch times using a two film model. The result is that the index for the outer oxide layer is $\tilde{N}_{f2} = 1.94$. Because of the similarity of the two indices, the thickness versus etch time curve is approximately correct as is, and therefore was not replotted.

C9.4   PMMA on InP

In 1984, Scheps[5] made a study of determining the film thickness and index of refraction of polymethyl methacrylate (PMMA) on InP. The particular emphasis is dealing with determining the index when the Δ/Ψ point falls near the period point. Substrates were iron doped <100> wafers, which were chemo-mechanically polished. The surface treatment consisted of a light scrub, a 3 min etch in 10 wt% HIO_3 followed by a rinse in deionized distilled water and blowing dry with dry N_2.

PMMA was deposited by spinning a solution of 9 wt% 496K PMMA in chlorobenzene for 60 s. The films were then baked for 75 min at 160°C. The film thickness was uniform within about 5%. Different thicknesses were formed by using different spin speeds, ranging from 2800 to 8000 rpm.

The ellipsometric data were taken with a manual ellipsometer using a wavelength of 6328 Å. The polarizer and analyzer readings were converted to Δ and Ψ. The concern addressed in this paper is how to deal with determining both the index and the thickness when the Δ/Ψ point falls near the period point.

The bare surface was measured and the corresponding index of refraction was determined to be $\tilde{N}_s = 3.47 - 0.45\,j$. These values might be contrasted to the values of Liu *et al.*[3] of $\tilde{N}_s = 3.521 - 0.300\,j$. This corresponds to a difference in Ψ of only about a half degree, but about 10° difference for Δ.

After deposition, the films were analyzed using angles of incidence of 30°, 50°, and 70°. The periodic film thicknesses for a film with index $\tilde{N}_f = 1.49$ for these angles of incidence are 2254, 2476, and 2736 Å, respectively. Generally, the separation of trajectories is greater for larger values of Ψ. Accordingly Scheps[5] used the data from the angle of incidence that gave the most separated trajectories to determine the

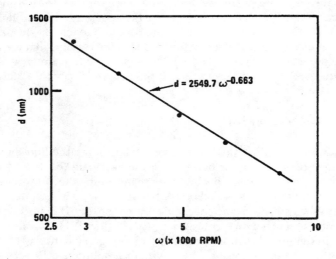

Figure C9-2. Film thickness as a function of terminal spin speed for a solution of 9% PMMA in chlorobenzene. Dots are data points, line is least-squares fit. (After Scheps[5])

index value and then used that index value with the other angles of incidence to determine thickness. Note that the author did not assume that all of the indices were the same. For film thicknesses far from the period point, the values of the index were determined to be 1.49 ± 0.1. When the points nearer the period point were included, the index ranged from 1.44 to 1.55.

The results of film thickness versus spin rate measurements are shown in Figure C9–2. An earlier work[12] attempted to fit data to an equation of the form

$$d = b\omega^n$$

where d is the film thickness in nm and ω is the angular velocity in 1000 rpm. This previous work[12] found the exponent to fall in the range of -0.4 to -0.7. The exponent for a steady-state Newtonian fluid is predicted[13] to be -2/3. A linear regression least-squares fit of the data collected in this work[5] results in an exponent of -0.663.

C9.5 References

1. D. L. Lile, *J. Vac. Sci. Technol. B*, **1**, 48 (1984).
2. A. Heller, B. Miller, and F. A. Thiel, *Appl. Phys. Lett.*, **38**, 282 (1981).
3. X. Liu, E. A. Irene, S. Hattangady, and G. Fountain , *J. . Electrochem. Soc.*, **137**, 2319 (1990).
4. X. Liu, J. W. Andrews, and E. A. Irene, *J. Electrochem. Soc.*, **138**, 1106, (1991).
5. R. Scheps, *J. . Electrochem. Soc.*, **131**, 540 (1984).
6. A. Gagnaire, J. Joseph, A. Etchebery, and J Gautron, *J. Electrochem. Soc.*, **132**, 1655 (1985).
7. R. Muller, *Appl. Phys. Lett.*, **57**, 1020 (1990).
8. G. P. Schwartz, W. A. Sunder, and J. E. Griffiths, *Appl. Phys. Lett.*, **37**, 925 (1980).
9. O. R. Monteiro and J. W. Evans, *J. Electrochem. Soc.*, **135**, 2366 (1988).
10. A. Nelson, K. Geib, and C. W. Wilmsen, *J. Appl. Phys.*, **54**, 4134 (1983).
11. E. Bergignat, G. Hollinger, and Y. Robach, *Surface Sci.*, **189/190**, 353 (1987).
12. P. O'Hagan and W. J. Daughton, Kodak Publication No. G-48, Rochester, NY (1978).
13. B. D. Washo, *IBM J. Res. Dev.*, **21**, 190 (1977).

Case 10:
Benzotriazole and Benzimidazole on Copper

C10.1 General

One method for inhibiting corrosion of copper is to apply a thin film of an azole such as benzotriazole (BTA) or benzimidazole (BIMDA).[1,2] The molecules are shown schematically in Figure C10-1. We have a benzine ring with one pair of carbon atoms which are also part of a five-member ring that contains nitrogen. For the BTA, the other three atoms of the five-member ring are nitrogen. For BIMDA, two of the others are N and the third is carbon (with attached hydrogen, not shown). The bonding with the copper occurs by substitution of a copper atom for the hydrogen that is attached to the nitrogen on both species.

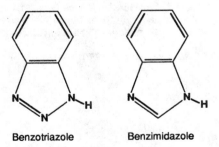

Benzotriazole   Benzimidazole

Figure C10-1. Schematic drawing of the benzotriazole (BTA) molecule and the benzimidazole (BIMDA) molecule.

The interaction of BTA and BIMDA with copper has been studied with a variety of analysis techniques including x-ray photoelectron spectroscopy,[3,4] infrared spectroscopy,[5,6] quartz crystal oscillator microbalance,[4] and ellipsometry.[7–10] In addition, electrochemical methods have been used by many. In this case study, we shall concentrate on the use of ellipsometry.

C10.2 *Ex situ* **Studies**

Hobbins and Roberts[7] have used ellipsometry to study the film growth of both BTA and BIMDA. The substrates used were vapor-deposited copper layers, about 1750 Å thick, deposited on optically polished silicon disks 2 in. in diameter. The substrates were degreased in boiling benzene for 5 min. A Cu_2O layer about 20 Å thick was grown[11] on the copper-coated substrates by immersing in dilute (1 vol%) aqueous nitric acid for 10 s and then rinsing in deionized distilled water. While the surface was still wet, the Cu-azole surface films were grown by immersing the samples in a stirred, warm (60°C) aqueous (0.017 M) solution of either BTA or BIMDA for a specified time. The samples were then rinsed in deionized distilled water. The samples were allowed to dry in the atmosphere and the ellipsometric measurements were made within 30 min. Ellipsometric measurements were made with a manual ellipsometer using wavelength 6328 Å and a 70° angle of incidence. The index of refraction used[12] for the substrate was $\tilde{N}_s = 0.272 - 3.24\,j$ and for the film was $\tilde{N}_f = 1.50$. The substrate index is representative of copper rather than copper oxide since it is known[13] that the Cu-azole surface film grows at the expense of the thin Cu_2O layer residing on the copper substrate. It should be noted that these authors[7] are using film-free values from others rather than measuring them from their own samples.

Figure C10-2 shows the results from these measurements. This figure suggests that the Cu-BTA film grows rapidly to a thickness of about 50 Å for very short immersion times (< 30 s). The growth rate is very slow after that and a thickness of about 100 Å is obtained for the 3 min immersion time. In contrast, the Cu-BIMDA film grows more rapidly to a thickness of about 175 to 425 Å, after which the growth is slow.

No film thicknesses were measured less than about 50 Å. Since the authors did not measure the film-free values with their samples, an incorrect starting point on the Δ/Ψ trajectory might be responsible for an

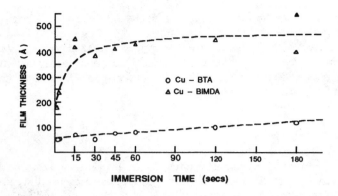

Figure C10-2. The thickness of Cu-BTA and Cu-BIMDA surface films as a function of immersion time. (After Hobbins[7])

offset of the entire plot. The differences in thicknesses of one film compared to another film should still be very accurate, however.

     It is interesting to compare these values with others. Tornkvist *et al.,*[6] using infrared spectroscopy, conclude that the BTA films are no more than 30 Å. Tompkins and Sharma[4] using a quartz crystal oscillator microbalance, conclude that similar films for BTA were 42 to 94 Å thick and that BIMDA films were from 490 to 2100 Å thick. The microbalance work[4] also shows that BIMDA in methanol etches the copper surface rather than forming a film.

C10.3 *In situ* Studies

     Cohen *et al.*[8] and Brusic *et al.*[9] in 1990 and 1991 report studies where they observe the formation of BTA films on Cu *in situ*. The oxidation state of the surface is controlled electrochemically. One of the questions Cohen addresses is the question as to whether BTA adsorbs on free Cu^0 surfaces in solution. The goal was to follow the adsorption of BTA under a wide range of conditions in a carefully controlled, well-characterized environment.

     The Cu surfaces were sputter-deposited Cu films approximately 2 μm thick on a thin layer of Ta on a Si wafer. The electrolytes were DI water, KOH, boric acid/borate buffer, nitric and sulfuric acids,

$(NH_4)_2S_2O_8$, and $Fe(NO_3)_3$, each alone or mixed with and without BTA and a surfactant. The *in situ* ellipsometry/electrochemistry experiment utilized an air-tight cell made of Kel-F with quartz windows. The cell was fitted with a Teflon "needle" through which a premeasured amount of BTA could be added.

Prior to adding BTA, several observations were made of the controlled oxidation and reduction of the Cu. Reproducible and controlled oxidation and reduction was observed in the pH range from 8 to 12. Some of the observations made at pH 12 are illustrated in Figure C10-3.

The sample was first simply immersed in the solution for 20 min. The value of Del stayed reasonably constant, indicating that the native oxide was stable in the solution. After application of a low cathodic potential (-1.6 V), Del increases by about 4.5° in about 11 s. This corresponds to the removal of the native oxide. After holding the potential here for 5 min, the potential was increased to -0.45 V and this

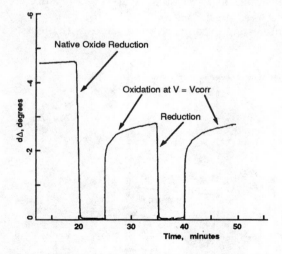

Figure C10-3. Ellipsometric parameter Δ as a function of time (in minutes) for Cu surface in N_2 saturated KOH at pH 12. Air formed oxide immersed in solution from 0 to 20 min; reduction at -1.6 V for 5 min; oxidation at -0.45 V from 25 to 35 min; reduction at -1.6 V again from 35 to 40 min; reoxidation from 40 to 50 min. (After Cohen[8])

leads to rapid growth of an oxide . This oxide can be reduced by returning the potential to -1.6 V. Figure C10-3 shows two cycles of the oxide reduction and growth.

Using x-ray photoelectron spectroscopy (XPS), in a related experiment, the authors[8] show that the film formed at -0.45 V is Cu_2O and that at -1.6 V, the surface is bare copper. They report the ellipsometer used, but do not report the wavelength of light. Based on the ellipsometer and the year the work was done, it seems reasonable to assume that the wavelength was 6328 Å. They conclude that the index of refraction of the oxide film was $\tilde{N}_f = 2.6 - 0.1j$. In this case, a change in film thickness of 5.5 Å causes a change in Δ of 1°. They conclude that the equilibrium thickness of the oxide grown at -0.45 V is about 14 Å.

In a similar manner, they show that CuO can be grown at pH 8.2 (boric acid/borate buffer) and higher, saturated with either N_2 or O_2 and with a potential of +0.2 V. The thickness of the CuO is typically 30 Å.

Based on the above, they show[8] that they can reproducibly produce a bare Cu surface, a thin Cu_2O surface oxide, or a thin CuO surface oxide. Many of their conclusions are substantiated with XPS. At this point, they introduce the BTA and the ellipsometry results are shown in Figure C10-4.

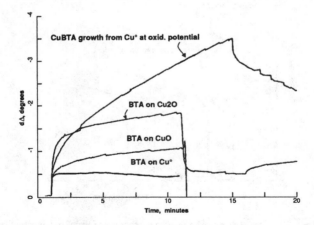

Figure C10-4. Ellipsometric parameter Δ as a function of time (in minutes) for uptake of BTA on Cu surfaces. Data are referenced to Δ value for BTA-free surfaces. Incomplete reduction of the BTA containing overlayer is observed at t = 12 min for Cu_2O and t = 15 min for the top curve. (After Cohen[8])

The thinnest film is formed on reduced Cu (-1.6 V). There is a rapid initial growth and a subsequent plateau at about 0.5°. The film that forms on the CuO (+0.2 V) is slightly thicker than the film that forms on the bare copper, and a still thicker film forms on Cu_2O. At about 12 min, a reducing potential is applied and most of the surface film is removed. Ancillary XPS analysis shows that the films are indeed adsorbed BTA.

The uppermost curve shows the film growth behavior when BTA is introduced to a bare copper surface (-1.6 V) and the potential is then increased to -0.45 V (the potential where Cu_2O grows in the absence of BTA). The film formed in this manner is somewhat less reducible than the film formed over the oxides.

The authors[8] conclude that the index of refraction for these films is $\tilde{N}_f = 1.8 - 0.05\,j$. In this case, a 1° change in Del represents 10.5 Å. This implies that the film which forms on metallic copper is about 5 Å thick. The film which forms on CuO is about 6 Å thick and grows to about 11 Å in 10 min. For the Cu_2O surface, the initial uptake is about 14 Å reaching about 20 Å after 10 min.

Based on the ellipsometry and XPS analysis, the authors conclude that films grown on reduced copper are bound to the Cu surface via electron donation from the triazole nitrogen molecular orbitals into the Cu conduction band. This layer is limited to one monolayer. At higher potentials where Cu ions are allowed to oxidize and dissolve, the ions are available to complex with the BTA in solution and therefore form more complex and thicker layers.

Summarizing this work, it is shown that reproducible well-controlled surfaces of reduced Cu, Cu_2O, and CuO can be formed and that BTA indeed will form a film on all of these surfaces.

A year later, Brusic *et al.*[9] from the same laboratory reported further experiments on the same system. Figure C10-5 shows a result similar to that shown in Figure C10-3 except that BTA is added in region VI and the additional film growth can be seen in the change in Δ. They observe that for pH values above 7, the oxide films can be reproducibly oxidized and reduced. The fact that the optical properties return to the values for oxide-free copper upon reduction at -1.6V is an indication that no appreciable surface roughening is occurring. Although the oxide can be reproducibly removed, note in Figure C10-5 that the Cu-BTA film cannot be fully reduced (region VII).

Figure C10-6 shows the result when various pH values are used. Without BTA, at a pH of 5 at the corrosion potential, large irreversible changes in Δ due to roughening of the surface are observed. In contrast,

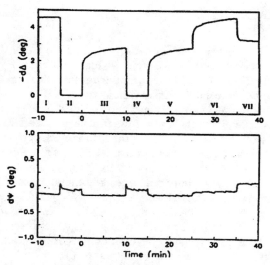

Figure C10-5. Variation of ellipsometric parameters Δ and Ψ with Cu surface reduction, reoxidation, and growth of Cu-BTA, in pH 12. (After Brusic[9])

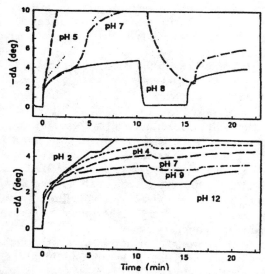

Figure C10-6. Variation of $d\Delta$ with spontaneous reoxidation of oxide-free Cu in solutions without 1H-BTA (upper graph) and with 1H-BTA (lower graph). Zero time corresponds to a release of the potentiostatic control at low potential, followed by oxidation at open-circuit potential. (After Brusic[9])

when BTA is present (starting from the bare surface) Cu oxidation at the corrosion potential is well behaved down to a pH of 4. At pH of 2, the ellipsometric data becomes irreproducible, probably due to surface roughening. One of the conclusions of this figure is that as the pH decreases, the film thickness increases.

Brusic *et al.*[9] observe that the growth kinetics are best represented by a logarithmic law at high pH and a parabolic law at neutral and mildly acidic pH. Figure C10-7 shows examples of the kinetics at a high and low pH.

The Cu-BTA growth kinetics also depend on whether the surface

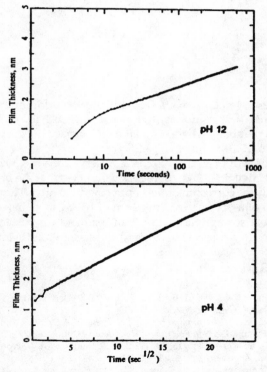

Figure C10-7. Kinetics of Cu-BTA film formation in pH 12 (upper graph) and pH 4 (lower graph). (After Brusic[9])

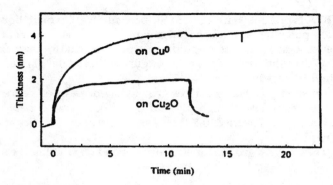

Figure C10-8. Difference in Cu-BTA film growth on oxidized and oxide-free Cu surface in pH 7. (After Brusic[9])

was oxide-covered or oxide-free prior to the addition of the BTA. In neutral (pH = 7) solutions, the growth of Cu-BTA on the oxide-free surface is parabolic whereas on an oxide surface the growth is logarithmic. Figure C10-8 shows this effect.

This work[9] goes on to make observations concerning the corrosion protection the BTA provides and observes that the presence of these layers reduces corrosion of the copper significantly. They conclude that a buildup of a polymerized network of BTA-Cu chains occurs and that this is most strongly favored on an oxidized Cu surface, in solutions where the Cu oxides are stable and Cu dissolution is slow.

C10.4   References

1. Proctor and Gamble Ltd. Britt. Pat. No. 652,339 (Dec. 1947).
2. R. Walker, *J. Chem. Ed.*, **57**, 789 (1980).
3. R. F. Roberts, *J. Electron Spectrosc. Relat. Phenom.*, **4**, 273, (1974); D. Chadwick and T. Hashemi, *J. Electron Spectrosc. Relat. Phenom.*, **10**, 79, (1977); D. Chadwick and T. Hashemi, *Corros. Sci.*, **18**, 39 (1978); A. R. Siedle, R. A. Velapoldi, and N. Erickson, *Appl. Surf. Sci.*, **3**, 299 (1979).
4. H. G. Tompkins and S. P. Sharma, *Surf. Interface Anal.*, **4**, 261 (1982).
5. H. G. Tompkins, D. L. Allara, and G. A. Pasteur, *Surf. Interface Anal.*, **5**, 101

(1983); G. W. Poling, *Corros. Sci.*, **10**, 359 (1970); S. Thiboult, *Corros. Sci.*, **17**, 701 (1977).

6. C. Tornkvist, D. Thierry, J. Bergman, B. Liedberg, and C. Leygraf, *J. Electrochem. Soc.*, **136**, 58 (1989).
7. N. D. Hobbins and R. F. Roberts, *Surface Technol.*, **9**, 235 (1979).
8. S. L. Cohen, V. S. Brusic, F. B. Kaufman, G. S. Frankel, S. Motakef, and B. Rush, *J. Vac. Sci. Technol. A*, **8**, 2417 (1990).
9. V. Brusic, M. A. Frisch, B. N. Eldridge, F. P. Novak, F. B. Kaufman, B. M. Rush, and G. S. Frankel, *J. Electrochem. Soc.*, **138**, 2253 (1991).
10. F. Mansfeld and T. Smith, *Corrosion*, **29**, 105 (1973).
11. J. Kurger, *J. Electrochem. Soc.*, **108**, 503 (1961).
12. F. Mansfeld, T. Smith and E. P. Parry, *Corrosion - NACE*, **27**, 289 (1971).
13. I. C. G. Ogle and G. W. Poling, *Can. Metall, Q.*, **14**, 37 (1971).

Case 11:
Adsorption on Metal Surfaces

C11.1 General

The use of ellipsometry for measuring very thin films was described in Chapter 5. We include here several additional examples of the use of ellipsometry for adsorption studies. Most adsorption studies are done *in situ* and as discussed in Chapter 5, oftentimes most of the information comes from Δ alone rather than both Δ and Ψ. The value of Δ is often not as important as $\delta\Delta$, the change in Δ. Note that in many of these studies, no mention is made of optical constants. The change in Δ is simply treated as a measure of film growth or adsorption.

C11.2 NO, O₂, and CO on Copper

Balkenende *et al.*[1,2] report a study of the interaction of NO, O₂, and CO with Cu(111), Cu(100), and Cu(110). The reason for this study is to understand the catalytic decomposition of NO for reducing pollution. Their studies on Cu(110) will illustrate the pertinent ellipsometric methods, so for simplicity, we will consider primarily their work on Cu(110).

The study was done using temperatures of 370 to 770 K and exposures up to 150 Pascal-seconds. The Cu single-crystal samples were spark-cut from high-purity copper rods and were cleaned using Ar ion sputtering and subsequent annealing at 720 K for 15 min, producing a clean and well-ordered surface. The ellipsometry measurements were made using an angle of incidence of 69° ± 1°, and a wavelength of 6328 Å. The quantity they record is $\delta\Delta$, which they show to be proportional to the total amount of oxygen and nitrogen present. In addition, they make measurements using Auger electron spectroscopy (AES) and low energy electron diffraction (LEED).

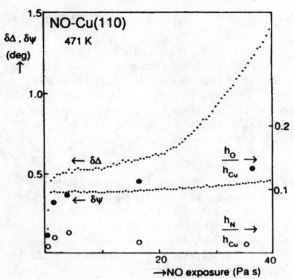

Figure C11-1. $\delta\Delta$, $\delta\Psi$, h_O/h_{Cu} and h_N/h_{Cu} versus exposure of NO on Cu(110) at 471 K. (After Balkenende[2])

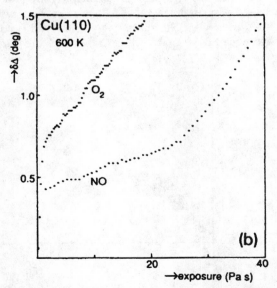

Figure C11-2. $\delta\Delta$ versus exposure of NO and O_2 on Cu(110) at 600 K. (After Balkenende[2])

Figure C11-1 shows the changes in Δ and Ψ for the adsorption of NO on Cu(110) at 471 K. Also shown are the Auger ratios, h_O/h_{Cu} and h_N/h_{Cu}. The adsorption of O_2 and that of NO is compared in Figure C11-2. With NO exposure, Δ changes rapidly until $\delta\Delta \approx 0.5°$ where the curve flattens out. The slope increases again when $\delta\Delta > 0.8°$. For the adsorption of O_2, a rapid change in Δ is observed up to $\delta\Delta \approx 0.8$. Upon further exposure, $\delta\Delta$ increases, but at a slower rate. When NO exposure up to $\delta\Delta \approx 0.5°$ is followed by O_2 exposure (not shown), a rapid increase to $\delta\Delta \approx 0.8°$ is observed.

No change is observed in $\delta\Delta$ upon evacuation of the gas phase, and electron irradiation during AES measurements did not change the ellipsometric parameters. This indicates that no molecularly adsorbed NO or O_2 is present[3] and that the adsorption is dissociative. Nuclear reaction analysis (NRA) established that $\delta\Delta$ varies linearly with the total amount of oxygen adsorbed at the surface or penetrated in subsurface layers.[3] In Figure C11-3 it is seen that the Auger ratio $(h_N + h_O)/h_{Cu}$ varies linearly with $\delta\Delta$ for submonolayer coverages. The slope is the same as that observed for h_O/h_{Cu} after O_2 exposure. This indicates that the sensitivity of $\delta\Delta$ is about equal for oxygen and nitrogen. The flattening of the curves is due to the fact that AES is more sensitive for the first layer than the second, etc. This flattening is observed to correspond to a surface coverage of $\theta \approx 0.4$, according to the NRA measurements.

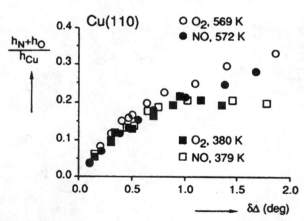

Figure C11-3. $(h_N + h_O)/h_{Cu}$ versus $\delta\Delta$ upon exposure of NO or O_2 at different temperatures for Cu(110). (After Balkenende[2])

AES measurements with Ar ion sputtering show that for coverages of 0.4 or less, the nitrogen and oxygen are removed simultaneously. For larger coverages, the oxygen takes longer to be sputtered away, implying that it is present in subsurface layers. No increase in the nitrogen surface coverage nor penetration of nitrogen atoms in subsurface layer was observed.

If oxygen is adsorbed on the surface and the surface is subsequently exposed to CO, the oxygen is removed from the surface, or reduced. This is shown in Figure C11-4 for adsorption on Cu(100). In the one case, only oxygen was adsorbed and the other case, both oxygen and nitrogen were adsorbed prior to the exposure to CO. The nitrogen coverages are indicated on the figure. When only oxygen was preadsorbed, the CO essentially removed all of the adsorbed species and Δ returned to the original value ($\delta\Delta = 0$). When some nitrogen was preadsorbed, however, the CO removes only the oxygen and leaves the adsorbed nitrogen. $\delta\Delta$ decreases to the value for the adsorbed nitrogen only.

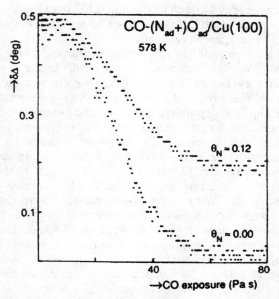

Figure C11-4. $\delta\Delta$ versus exposure of CO to Cu(100) previously exposed to O$_2$ ($\theta_N \approx$ 0.00) or NO ($\theta_N \approx 0.12$) up to $\delta\Delta \approx 0.5°$. (After Balkenende[2])

C11.3 Oxidation of Nickel

Deckers *et al*.[4] have studied the oxidation mechanism of clean and Pt-covered Ni(111) at high temperatures. Both platinum and nickel are important catalysts, often used in hydrogenation reactions.[5] In contrast to nickel, platinum is a very active oxidation catalyst due to the sub-monolayer chemisorption of atomic oxygen without the occurrence of further oxidation.[6] Ni, on the other hand, reacts with oxygen to form a surface layer of NiO. This study[4] focuses on the oxidation mechanisms of clean and Pt-covered Ni(111) surfaces at high temperatures.

Experimental

Ellipsometry is used to monitor the oxidation processes. Nuclear reaction analysis (NRA) is used for determining the absolute amounts of oxygen present. Auger electron spectroscopy (AES) is used along with argon ion sputtering to monitor the depth distribution of the dissolved oxygen. Low energy electron diffraction (LEED) and Rutherford back-scattering spectroscopy (RBS) are used to monitor the amount and structure of the platinum coverages.

The nickel crystal was spark-cut to within $0.2°$ of the (111) orientation, then ground and mechanically polished. After many sputtering and annealing cycles, a sharp (1x1) LEED pattern was observed. No contamination was observed with AES. In between the oxygen exposures, the sample was sputtered overnight and annealed at $550°C$. This treatment was sufficient to remove all traces of oxygen from the sample.

The oxygen exposures are expressed in Langmuir units (1 L = 10^{-6} Torr seconds) with the maximum exposure being 2500 L. The exposure was done at an O_2 pressure of 5×10^{-7} Torr, measured with a quadrupole mass spectrometer. Immediately after exposure, the sample was allowed to cool down for the various measurements. Ellipsometric off-null intensity measurements were made using a wavelength of 5540 Å. It has been shown[7] with AES that $\delta\Delta$ is approximately linearly related to the thickness of the oxide layer. The authors of Ref. 4 state "The presence of chemisorbed oxygen is not revealed in these ellipsometric measurements." This is in clear contradiction to the conclusions of the previous section and the following section. They later state "It appears that oxygen atoms present in the surface region and in the subsurface region of the crystal do not give rise to a different ellipsometric response." We shall not dwell on this question since neither their observations nor their conclusions depend strongly on this matter.

Results

The oxidation behavior of clean Ni(111) at different temperatures is shown in Figure C11-5. The room temperature curve shows, after an initial incubation period, a rapid increase up to a saturation value of $\delta\Delta = 2.1°$. This marks the formation of a closed NiO layer (a few atoms deep) on top of the Ni(111) surface and represents about 3.4×10^{15} O_2 atoms/cm^2. At 150°C, the behavior is similar except that no saturation is observed. At 250°C and 400°C, no incubation period is seen and the oxygen penetrates much deeper into the material.

Platinum was deposited on the clean Ni surface. The amount of Pt is listed as the number of monolayer equivalents (MLE). One MLE of deposited Pt atoms is defined as the number of atoms in a Ni(111) layer, i.e., 1.86×10^{15} atoms/cm^2. The effect of depositing submonolayer coverages of Pt on the Ni surface on the oxygen uptake at room temperature is shown in Figure C11-6. Pt suppresses the room temperature

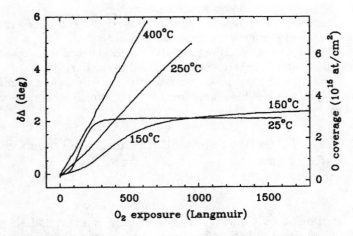

Figure C11-5. $\delta\Delta$ and absolute oxygen coverage as a function of $^{18}O_2$ exposure to a Ni(111) surface at various temperatures. (After Deckers[4])

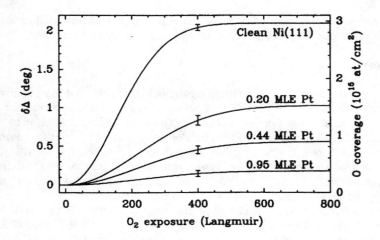

Figure C11-6. Ellipsometric oxygen uptake curve at room temperature of Ni(111) with different platinum coverages as measured separately using RBS. (After Deckers[4])

oxidation of Ni(111). Note that 0.2 monolayers of Pt reduced the oxidation to 50% of the saturation value for pure Ni.

Figure C11-7 shows the effect for adsorption at higher temperatures. At the higher temperatures, the Pt and Ni have formed an alloy layer. The oxygen interacts with this layer and penetrates into the bulk, although at a much reduced rate compared to the pure Ni.

The paper discusses various other aspects of the oxidation mechanism, but the use of ellipsometry for this application has been described sufficiently.

C11.4 The Interaction of Small Amounts of Oxygen with Aluminum Films

General

Grimblot and Eldridge[8] report a study of the initial stages of oxidation of aluminum thin films. A previous study[9] of oxygen interaction with single-crystal Al is described in Chapter 5. Work function

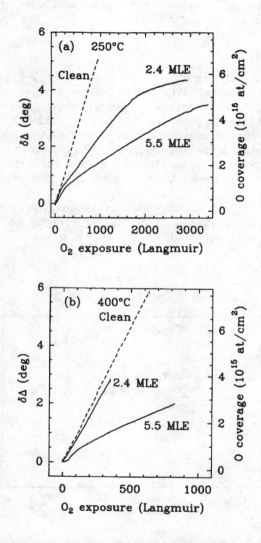

Figure C11-7. Ellipsometric oxygen uptake curves of Ni(111), pure and with various amounts of platinum, oxidized at (a) 250°C and (b) 400°C. (After Deckers[4])

studies[10] suggest that the oxygen is directly incorporated into an oxide film, although XPS measurements suggest that some simple chemisorbed species can exist on Al(111).

Experimental

2000 Å aluminum films were deposited using E-gun evaporation in a large UHV system. The Al was outgassed and deposited over a large area of the UHV chamber to getter reactive gases prior to film deposition at a system pressure < 10^{-8} Torr. Residual gas analysis was used to measure the vacuum system performance, but the use of hot filaments was minimized during the oxygen exposure in order to minimize the formation of activated oxygen species. An ion pump was kept on during the exposures to estimate the oxygen pressures from pump currents.

An *in situ* manual ellipsometer was used with an angle of incidence of 70° and a wavelength of 5461 Å. Roughly 1 to 2 min were required to determine Δ and Ψ.

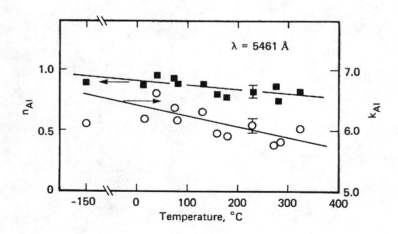

Figure C11-8. Variation of the refractive index and extinction coefficient of Al(111) films with temperature at a wavelength of 5461 Å. (After Grimblot [8])

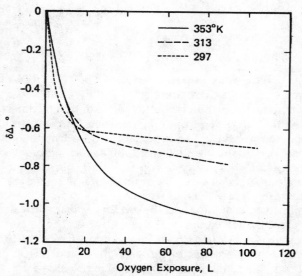

Figure C11-9. Dependence of δΔ on O_2 exposure and temperature. (After Grimblot[8])

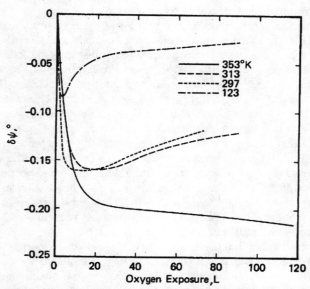

Figure C11-10. Dependence of δΨ on O_2 exposure and temperature. (After Grimblot[8])

Results

The Al films were deposited at about 300 K. SEM and x-ray studies showed that the films were smooth with essentially (111) preferred orientation and a grain size comparable to film thickness. Figure C11-8 shows the values of the optical parameters for the bare surfaces as a function of temperature. The authors of Ref. 8 attribute the temperature dependence to small undetectable hillock growth and/or minor densification effects. They conclude that their film resembles clean Al(111) single-crystal material. The grain boundaries were estimated to constitute roughly 1% of the film surface.

Typical variation of Δ and Ψ with O_2 exposure at various temperatures are shown in Figures C11-9 and C11-10. The $\delta\Delta$ curve for the 123 K exposure is not shown since it superimposes upon the 313 K data. The $\delta\Delta$ data show decreases in the value of Δ and this implies film growth of some form, either as chemisorption or oxide growth. If this were simply the increase in thickness of a single film, based on the discussions of Δ/Ψ trajectories in Chapter 3, the value of Ψ would be expected to also increase. The $\delta\Psi$ versus exposure data in Figure C11-10 clearly show an initial decrease followed by an increase in Ψ. The minima broaden and shift to the right with increasing temperature.

The $\delta\Psi$ minima appear to define the approximate completion of the regime where oxygen incorporation rates significantly exceed chemisorption. This contention is supported by surface potential measurements of Hofmann.[10] Hofmann also reports that increasing the O_2 pressure at constant temperature accelerated the oxygen chemisorption relative to incorporation. Grimblot and Eldridge[8] also observe this. In addition, with constant pressure, one can deduce from Figures C11-9 and C11-10 that raising the temperature promotes incorporation over chemisorption. The 353 K exposure did not achieve the minima in $\delta\Psi$.

Prior to the $\delta\Psi$ minima, the average thickness is proportional to $L^{1/2}$ whereas after the minima, the growth is logarithmic. AES and XPS studies[11] show that no Al^{+3} signal is present at low exposures. Reference 8 concludes that the film formed by incorporation is not true Al_2O_3 since this does not form until much higher exposures.

It is concluded that the interaction of oxygen at low pressure with clean (111) oriented Al films is complex. The observations are compatible with the view that a surface reconstruction or place exchange competes with a chemisorption process.

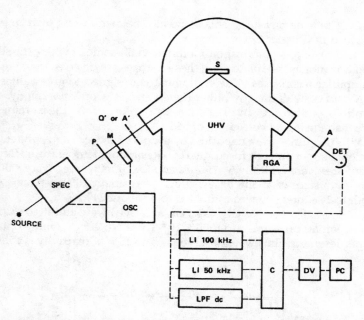

Figure C11-11. Ellipsometer schematic. The components are labeled as follows: spectrometer (SPEC), linear polarizer (P), modulator (M), modulator driver unit (OSC), calibration circular/linear analyzer (Q' or A'), sample (S) in ultrahigh vacuum chamber with residual gas analyzer (RGA), linear analyzer (A), photomultiplier (DET), lock-in detectors (LI 100 kHz, LI 50 kHz), low pass filter (LPF), coupler/converter (C), digital voltmeter (DV), and programmable calculator (PC). (After O'Handley[12])

C11.5 Residual Gases on Silver

O'Handley *et al.*[12] reports a study on the effects on the optical properties of silver films by specific residual gases present during deposition. The schematic of the experimental setup is shown in Figure C11-11. The deposition is done in an ultrahigh vacuum system and the ellipsometry is done with a specially designed ellipsometer which emphasizes the measurements of small changes in Δ and Ψ. The use of this type of ellipsometer makes the measurements easier, faster, and more accurate, but the measurements could also be made using instruments described in Chapter 2.

They used a polarization-modulated ellipsometer where the state of polarization is modulated and these changes are detected using lock-in amplifier techniques. The equipment is designed to give values of $\sin(\Delta)$ and $\cos(2\Psi)$ directly rather than Δ and Ψ. Correspondingly, the quantities plotted are $\sin(\Delta)$ and $\cos(2\Psi)$. Most of the measurements were made using a wavelength of 5461 Å.

Silver films were deposited by evaporation from a molybdenum boat onto super-smooth fused-quartz flats at room temperature. Deposition rates were typically 5 to 9 Å/s and the films were polycrystalline with grain sizes of about 200 Å, which were about 1000 Å thick. The depth to $1/e$ intensity was estimated to be about 240 Å.

Several gases (N_2, CO, CO_2, H_2, and CH_4) were admitted during deposition at pressures up to 6×10^{-6} Torr. No effects on the optical properties of the films were observed. This lack of reactivity is consis-

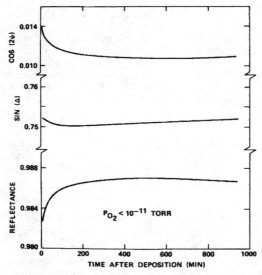

Figure C11-12. Measured ellipsometric parameters, $\cos(2\Psi)$ and $\sin(\Delta)$, and derived reflectance, R, as a function of time for a film deposited and maintained in UHV. Measurements are for λ = 5461 Å, angle of incidence = 63.9°. (After O'Handley[12])

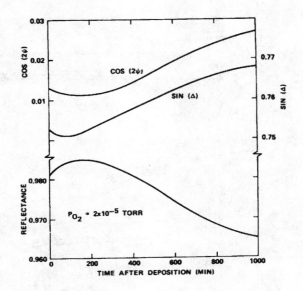

Figure C11-13. Same parameters as Figure C11-12, but for a film deposited in an oxygen partial pressure of 2×10^{-5} Torr. The leak was closed immediately after deposition. (After O'Handley[12])

tent with the results of others.[13] The presence of O_2 and H_2O during deposition significantly reduced the reflectivity when the partial pressures of these gases were above 10^{-6} Torr.

Figure C11-12 shows the changes in the optical parameters during room temperature annealing. The deposition and room temperature anneal were done with a partial pressure of O_2 less than 10^{-11} Torr. Figure C11-13 shows similar measurements for a film deposited in an oxygen partial pressure of 2×10^{-5} Torr, where the oxygen was pumped out immediately after deposition. In a similar experiment where the oxygen was left in place in the vacuum system, the changes were significantly less, being more like those shown in Figure C11-12. The changes shown in Figure C11-13 represent changes in Δ of about 2.3° and changes in Ψ of about 0.43°.

The authors of Ref. 12 take the early (< 4 h) changes to imply grain growth within the film along with surface smoothing of the surface

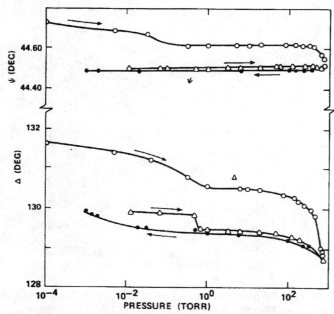

Figure C11-14. Ellipsometric parameters, Ψ and Δ, as a function of the pressure for a typical stabilized silver film. The film was cycled from high vacuum to air (open circles), back to vacuum (solid circles), and finally back to air (triangles). Parameters measured at $\lambda = 5461$ Å and angle of incidence = 63.9°. (After O'Handley[12])

rather than film growth, with the adsorbed oxygen affecting the surface diffusion. The later changes are assigned to the adsorption of residual water in the vacuum system.

Figure C11-14 shows the changes in Δ and Ψ when these films are cycled from vacuum to room air and back to vacuum. When first brought to atmospheric pressure, Δ and Ψ undergo large, step-like irreversible changes at about 0.1 Torr. On further pressure cycling, the changes in Ψ are in the opposite direction and smaller than before, and are reversible. The changes in Δ are similar to the initial irreversible changes, but are now smaller and reversible.

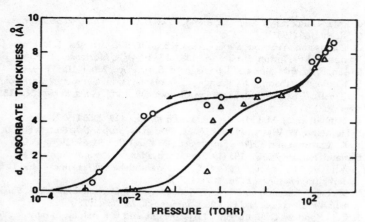

Figure C11-15. Surface film thickness of adsorbed water on silver as a function of pressure. Thicknesses derived from reversible portion of curves in Figure C11-14 assuming the index of the layer to be 1.33. Solid curves represent BET theory (see text). (After O'Handley[12])

The irreversible changes[12] are assigned to oxygen chemisorption on the silver surface and the reversible changes are assigned to the adsorption of water. If one assumes an index for adsorbed water to be 1.33, the step changes in Figure C11-14 give a monolayer thickness of 5.5 Å and a thickness at atmospheric pressure of about 20 Å. Figure C11-15 shows the resulting adsorption isotherm. This is the shape predicted by the BET theory.[14]

C11.6 References

1. A. R. Balkenende, O. L. J. Gijzeman, and J. W. Geus, *Appl. Surf. Sci.*, **37**, 189 (1989).
2. A. R. Balkenende, H. den Daas, M. Huisman, O. L. J. Gijzeman, and J. W. Geus, *Appl. Surf. Sci.*, **47**, 341 (1991).
3. A. R. Balkenende, G. W. R. Leibbrandt, M. Wiegel, O. L. J. Gijzeman, F. P. H. M. Habraken, and J. W. Geus, to be published.
4. S. Deckers, F. H. P. M. Habraken, W. F. van der Weg, and J. W. Geus, *Appl. Surf. Sci.*, **45**, 207 (1990).

5. C. Rofer-DePoorter, *Chem. Rev.*, **81**, 447 (1981); P. Biloen and W. Sachtler, *Adv. Catal.*, **30**, 165 (1981)

6   D. Collins and W. Spicer, *Surface Sci.*, **69**, 85 (1977); K. Griffiths, T. Jackman, J. Davies, and P. Norton, *Surface Sci.*, **138**, 113 (1984); K. Mortensen, C. Klink, F. Jensen, F. Besenbacher, and I. Stensgaard, *Surface Sci.*, **220**, L701 (1989).

7. P. deBokx, F. Labohm, O. Gijzeman, G. Bootsma, and J. Geus, *Appl. Surf. Sci.*, **5**, 321 (1980); F. Labohm, O. Gijzeman, G. Bootsma, and J. Geus, *Surface Sci.*, **135**, 409 (1983).

8. J. Grimblot and J. M. Eldridge, *J. Electrochem. Soc.*, **129**, 2366 (1982).

9. P. Hofmann, W. Wyrobisch, and A. M. Bradshaw, *Surface Sci.*, **80**, 344 (1979).

10. V. K. Agarwala and T. Fort, *Surface Sci.*, **45**, 470 (1974), **54**, 60 (1976); P. O. Garland, *Surface Sci.*, **62**, 183 (1977); A. M. Bradshaw, P. Hofmann, and W. Wyrobisch, *Surface Sci.*, **68**, 269 (1977); A. M. Bradshaw, W. Domcke, and L. S. Cederbaum, *Phys. Rev. B*, **16**, 1480 (1977).

11. R. Michel, J. Gastaldi, C. Allasia, C. Jourdan, and J. Derrien, *Surface Sci.*, **95**, 309 (1980); W. Eberhardt and C. Kunz, *Surface Sci.*, **75**, 709 (1978).

12. R. C. O'Handley, D. K. Burge, S. N. Jasperson, and E. J. Ashley, *Surface Sci.*, **50**, 407 (1975).

13. D. O. Hayward, in "Chemisorption and Reactions on Metallic Films," Vol. 1, edited by J. R. Anderson, Academic Press, New York (1971) p. 259.

14. S. Brunauer, P. H. Emmett, and E. Teller, *J. Am. Chem. Soc.*, **60**, 309 (1938); A. W. Adamson, "Physical Chemistry of Surfaces," 2nd ed., Interscience, New York (1967) p. 584.

Case 12:
Silicon-Germanium Thin Films

C12.1 General

$Si_{1-x}Ge_x$ layers deposited on silicon substrates are of interest for the fabrication of advanced electronic devices. Racanelli *et al.*[1] reported a study where ellipsometry was used to determine the thickness and stoichiometry of the layer.

C12.2 Experimental

$Si_{1-x}Ge_x$ layers were deposited onto silicon substrates by chemical vapor deposition in a commercial, atmospheric pressure reactor at 625°C from SiH_2Cl_2 and GeH_4 in H_2.[2] Two to ten films with different thicknesses were grown for each of six different germanium fractions spanning the range of $x = 0.10$ to $x = 0.30$. The ellipsometry was done using an angle of incidence of 70° and a wavelength of 6328 Å. For comparison in setting up the method, thickness and composition were determined on selected samples by cross-sectional transmission electron microscopy (XTEM) and Rutherford backscattering spectrometry (RBS).

C12.3 Results

Figure C12-1 shows the Δ/Ψ values obtained for three different stoichiometries. For the calculation, the substrate index is taken as $\tilde{N} = 3.858 - 0.018j$. It is assumed that there is a 12 Å native oxide on the Si with index $\tilde{N} = 1.46 - 0.0j$. This causes the starting point for the Si-Ge film to be $175.80°/10.41°$. This is called the "film-free" point.

213

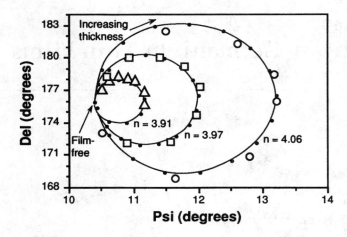

Figure C12-1. Measured points and calculated curves for Δ and Ψ for $x = 0.10$ (triangles), $x = 0.19$ (squares), and $x = 0.31$ (circles) for $Si_{1-x}Ge_x$ thicknesses ranging from 0 to 1000 Å. The solid dots are at 50 Å intervals. (After Racanelli[1])

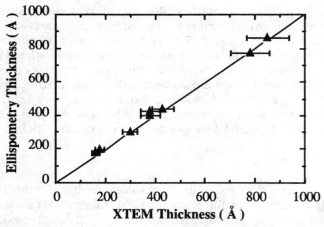

Figure C12-2. Comparison of thickness measured by ellipsometry and XTEM. The line drawn is the expected response for perfect agreement. (After Racanelli[1])

The solid lines are the calculated curves using the value of n shown in the figure and k = 0.018. Because k is nonzero, the curves do not quite close on themselves, although the offset is quite small. Only one cycle is shown. The solid dots represent 50 Å intervals on the calculated curves. We can use these curves to determine thickness by inspection and Figure C12-2 shows the correlation between the thickness determined in this manner and the thickness as determined by XTEM measurements. Figure C12-3 shows the correlation between the real part of the index of refraction and the germanium concentration as determined by RBS. Also shown are the results of Lukes et al.[3] measured by reflectivity on bulk $Si_{1-x}Ge_x$ samples. The real part of the index of refraction can be written approximately as

$$n(x) = 3.857 + 0.502\,x + 0.521\,x^2$$

while the imaginary part appears to remain constant at k = 0.018.

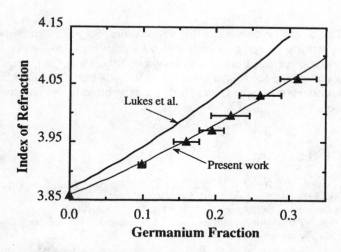

Figure C12-3. Real part of the index of refraction of $Si_{1-x}Ge_x$ as a function of germanium fraction. Results are plotted along with those of Lukes et al.[3] measured by reflectivity on bulk $Si_{1-x}Ge_x$ samples. In the present work[1], the index of refraction is determined by ellipsometry and the germainum fraction by RBS. (After Racanelli[1])

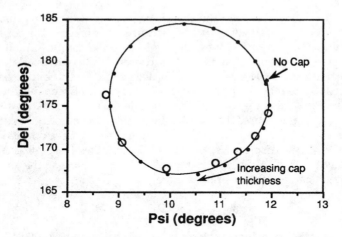

Figure C12-4. Comparison of measured points and calculated curves for Δ and Ψ for a nominally 350 Å $Si_{0.8}Ge_{0.2}$ layer capped by a silicon layer of varying thickness. The solid dots represent 50 Å intervals on the calculated curves in cap layer thickness in the range of 0 to 1000 Å. (After Racanelli[1])

For heterojunction transistor applications, $Si_{1-x}Ge_x$ films capped by a Si layer are of interest. Figure C12-4 shows measured and calculated data for varying thickness Si layers on top of a Si-Ge layer. The index used for the silicon was the same as for single crystal silicon. The fit to the curve is quite good, considering that no other fitting parameters were used.

C12.4 References

1. M. Racanelli, C. I. Drowley, N. D. Theodore, R. B. Gregory, H. G. Tompkins, and D. J.Meyer, *Appl. Phys. Lett.*, **60**, 2225 (1992).
2. W. B. deBoer and D.J. Meyer, *Appl. Phys. Lett.*, **58**, 1286 (1991).
3. F. Lukes, W. Schnmidt, J. Humlicek, M. K. Kekoua, and E. Khoutsichvili, *Phys. Stat. Sol.*, **53**, 321 (1979).

Case 13:
Profiling of HgCdTe

C13.1 General

HgCdTe heterostructures are used for IR detectors and are grown epitaxially, using molecular beam epitaxy (MBE), metalorganic chemical vapor deposition (MOCVD), or liquid-phase epitaxy (LPE). McLevige et al.[1] address the need for a post-growth characterization tool for material screening prior to device processing.

Rhiger and Kvaas[2] have reported previously that the ellipsometric parameter Ψ varies directly with x for "clean" $Hg_{1-x}Cd_xTe$ over a range of composition of x = 0.2 to 0.35. The value of Δ typically remains in the region of 146° to 148°.

C13.2 Experimental

McLevige et al.[1] obtain a "profile" by chemically etching the sample and measuring the ellipsometric parameters. The measured parameters depend on the etch method. They comment that for successive step etching, only the bromine/methanol and bromine/ethylene glycol etches were found to give consistently reliable surfaces for determining the ellipsometric parameters. Because bromine-based etches age quickly, fresh batches of etch were used each time.

The heterostructure $Hg_{1-x}Cd_xTe$ layers used in the study were grown by the three methods listed previously, namely, LPE, MOCVD, and MBE. The procedure was to measure the starting thickness of the sample by cross-sectional methods, plot the ellipsometric parameters as a function of etch time. The transition to the CdTe buffer layer was later calibrated to etch time.

C13.3 Results

For calibration purposes, the composition of several different heterostructures was determined using the diode cutoff wavelength method.[1] The ellipsometric parameters Δ and Ψ were also measured. Figure C13-1 shows a plot of the x composition of $Hg_{1-x}Cd_xTe$ as a function of the parameter Ψ. Included are data from Rhiger and Kvaas[2] and Arwin et al.[3] The values of Δ remained in the 146° to 148° range. The solid line represents a quadratic fit to the data. The fit also included measured values for x = 0, (HgTe) with $\Psi = 14.7°$ and x = 1, (CdTe) with $\Psi = 4.5°$, although these points are not shown in the figure.

One sample was cross-sectioned and measured with x-ray energy dispersive spectroscopy (called "EDAX" in Figure C13-2). A sample of this material was also etch profiled with ellipsometry. The resulting profile along with the cross-sectional EDAX measurement is shown in Figure C13-2. Although there is some discrepancy in the depth scale, the agreement is well within the accepted accuracy of the energy dispersive technique.

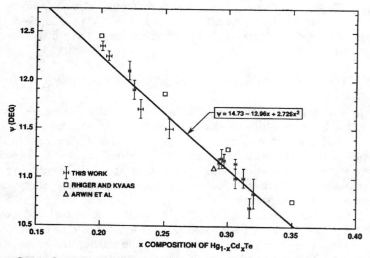

Figure C13-1. Curve fit of ellipsometric Ψ parameter data as a function of x composition of $Hg_{1-x}Cd_xTe$. Also plotted are data from Rhiger and Kvaas[2] and Arwin et al.[3] (After McLevige[1])

Another sample, which consisted of a 1.5 µm CdTe cap layer on top of a 16 µm HgCdTe active layer on top of a 6 µm CdTe buffer layer on a GaAs substrate, was measured by ellipsometry along with etching and the resulting profile is shown in Figure C13-3. The CdTe layers are very evident from the Ψ plot. Δ is nearly constant at about 147° in the HgCdTe layer but dips down to 134° to 142° in the CdTe layers. This lower value is consistent with other works.[2,3]

C13.4 Discussion

The authors do not address the question of the depth sampled by this method. We can readily calculate this, however, from the measured values of Δ and Ψ. With Δ/Ψ approximately 148°/12°, the program in Appendix A or the plot in Figure 3-4 can be used to determine that the extinction coefficient is approximately equal to 0.823 and the depth to 1/e for this value of k is 612 Å. The sampled depth would then be roughly three times this or about 1800 Å. With this in mind, we conclude

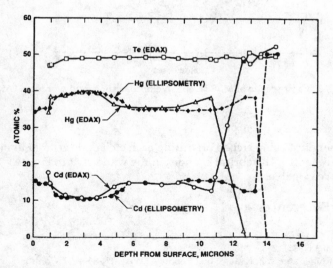

Figure C13-2. Energy dispersive x-ray spectroscopy ("EDAX") and ellipsometrically determined atomic composition profiles of MOCVD layer 4-248. (After McLevige[1])

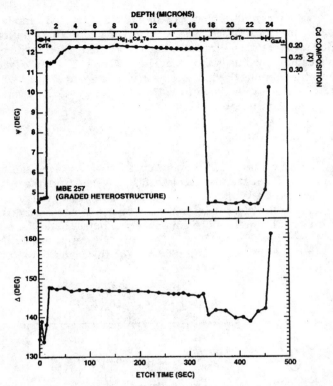

Figure C13-3. Cross section of MBE layer 257, ellipsometric Ψ and Δ profiles. (After McLevige[1])

that the amount being etched off during each etch cycle is more than the measured depth. The abrupt changes in the value of Ψ in Figure C13-3 are then reasonable.

C13.5 References

1. W. V. McLevige, J. M. Arias, D. D. Edwall, and S. L. Johnston, *J. Vac. Sci. Technol. B*, **9**, 2483 (1991).
2. D. R. Rhiger and R. E. Kvaas, Final Report AFWAL-TR-86-4009, Contract F33615-80-C-5084 (1986).
3. H. Arwin, D. E. Aspnes, and D. R. Rhiger, *J. Appl. Phys.*, **54**, 7132 (1983).

Case 14:
Oxides and Nitrides of Silicon

C14.1 General

The measurement of the thickness of thermal oxide or low pressure chemical vapor deposition nitride on silicon is probably the classic turnkey ellipsometric operation. Many ellipsometers have the optical constants for thermal oxide already included in the software. In this case study, we consider some variations of these measurements. We start with a study of the effect of using lower temperature to grow the thermal oxide. This is followed by a study of plasma silicon nitride. Lastly, a study of silicon oxynitride is presented.

C14.2 Low Temperature Growth of Silicon Dioxide

Lucovsky et al.[1] report a study of the effect of using lower than normal temperatures to oxidize silicon. Various references given in this work indicate that film and interface quality are generally degraded and that oxide density and intrinsic stress are observed to increase substantially when oxidation temperatures are reduced below 1000°C.

Ellipsometry and infrared spectroscopy (IR) are combined to give information about chemical bonding. The silicon substrates used were (111) single-crystals, which were polished on both sides. They were cleaned using the RCA method,[2] followed by an HF dip and a thorough rinse in deionized water. They were then blown dry using clean N_2. Oxidation was done using pure dry oxygen at pressures between 1 and 300 atm and at temperatures between 550 and 1000°C.

From an ellipsometric point of view, the primary interest was in the index of refraction. Accordingly, the oxide thickness was grown to about one-half of an ellipsometric period where the accuracy of the index values is greatest. The ellipsometric measurements were made using a wavelength of 6328 Å. The IR measurements were made using a dispersive double beam instrument. The IR absorption peak of interest is the asymmetric oxygen stretching peak that occurs in the region of 1070 cm^{-1} (1070 wavenumbers).

Figure C14-1 shows the IR spectra for several different oxidation conditions. Clearly, the peak position depends on the oxide growth temperature. Figure C14-2 shows a plot of these variations along with a plot of the index of refraction measured with ellipsometry. Lucovsky *et al.*[1] make the observation that correlations between the index of refraction and the IR stretching frequency at higher pressures are essentially the same as at 1 atm.

Figure C14-3 shows the correlation between index and stretching frequency. After discussion of various mechanisms causing these changes

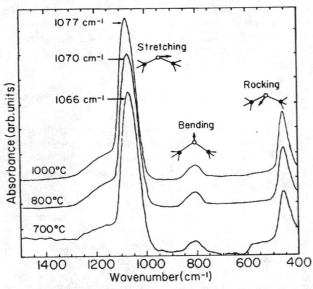

Figure C14-1. Normalized IR absorption vs. wave number for three films grown at 1 atm of dry oxygen at temperatures of 700, 800, and 1000°C. The atomic displacements of the three features are also indicated. (After Lucovsky[1])

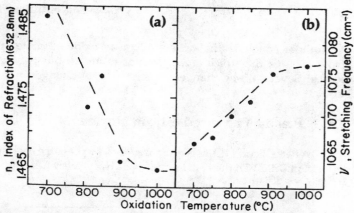

Figure C14-2. (a) Index of refraction at 6328 Å vs. oxidation temperature for films grown at 1 atm of dry oxygen. (b) IR stretching vibration wavenumber vs. oxidation temperature for films grown at 1 atm of dry oxygen. The dashed lines indicate the trends in the data. (After Lucovsky[1])

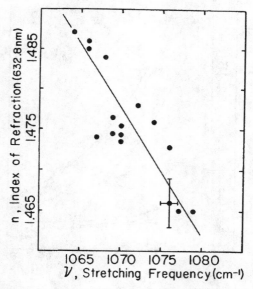

Figure C14-3. Index of refraction vs. frequency of IR stretching vibration for all low-temperature oxide films. The experimental data are for all of the oxides grown and the solid line is derived from model calculation. The degree of uncertainty is shown by the error bars shown on one of the experimental points. (After Lucovsky[1])

the authors conclude that when silicon is oxidized at temperatures below 1000°C, the stoichiometry is not changed, but the density increases. Several of the changes are explained as changes in the Si-O-Si bond angle.

C14.3 Plasma Enhanced Silicon Nitride

When silicon nitride is prepared by low pressure chemical vapor deposition (LPCVD) the stoichiometry is always very near Si_3N_4. When

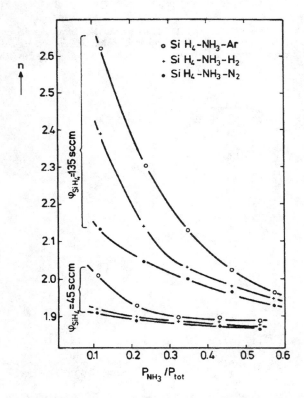

Figure C14-4. The refractive index of plasma silicon nitride layers deposited at 300°C, a total pressure of 0.33 mbar and an RF power of 200 W, as a function of the NH_3 partial pressure for the SiH_4-NH_3-x system (x = Ar, N_2, H_2). (After Claassen[3])

silicon nitride is prepared by plasma enhanced chemical vapor deposition (PECVD), the process adds hydrogen to the material and the stoichiometry can vary considerably. Claassen et al.[3] report a study of the characterization of PECVD silicon nitride as a function of the processing conditions.

Ellipsometry was used for index of refraction measurements, Auger electron spectroscopy (AES) and Rutherford backscattering spectrometry (RBS) were used for measurements of the Si/N ratio, and infrared spectroscopy (IR) was used for the hydrogen concentration measurements.

A low-pressure, radial-flow parallel-plate reactor operating at 50 kHz was used for depositing the samples. Nitride layers were deposited onto silicon wafers at a substrate temperature of 300°C with an RF power of 200 W. The total pressure was 0.33 mbar. Nitrides with different compositions were deposited by varying the flows of the input gases, SiH_4, NH_3, and N_2. Some layers were also deposited using H_2 or Ar instead of N_2.

Figure C14-4 shows the refractive index n (note that k = zero for all of these films) plotted as a function of the gas-phase composition during deposition. The refractive index increases when the amount of NH_3 is reduced and this effect is more pronounced for the higher SiH_4 flow. As will be discussed below, a high index indicates a nitrogen-poor film. N_2 is harder to dissociate than is NH_3, therefore decreasing the NH_3/N_2 ratio leads to a nitrogen-poor film.

Figure C14-5 shows the relationship between refractive index as measured by ellipsometry and the Si/N ratio, as measured by RBS. The refractive index varies linearly as a function of the Si/N ratio. We see from this figure that for a film with Si/N ratio of 0.75, the index is 1.9. An LPCVD film will have this ratio, but the index will normally be 2.0. The difference is due to the large amount of hydrogen incorporated in the plasma nitride films.

Hydrogen is bound as Si-H and N-H in these films. Using a method of Lanford and Rand,[4] the amount of each of these species can be determined using IR, and Claassen et al.[3] show that for all of the films in this study deposited with NH_3, the percent hydrogen remains constant. Chow et al.[5] measured the hydrogen content of PECVD nitride films using nuclear reaction analysis and concluded that the concentration of hydrogen in these layers varies as a function of the deposition temperature. They found that at 300°C about 23.7% of the film was hydrogen. This is in good agreement with Claassen et al.[3]

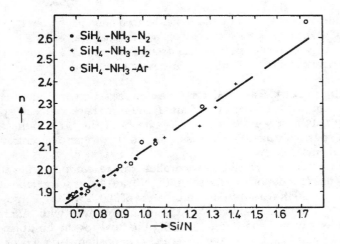

C14-5. The refractive index of plasma silicon nitride layers deposited at 300°C as a function of the Si/N ratio as measured by RBS. (After Claassen[3])

Figure C14-6 shows the relationship between the index of refraction and the Si-H/Si-N bonds as measured with IR. This relationship is also linear, which implies that the ratio of Si-H/N-H bonds is directly proportional to the atomic ratio, Si/N.

Some experiments were done in a home-built furnace-type parallel-plate reactor at operating frequencies of 400 kHz and 3 MHz at 300°C and 500°C. The resulting Si-H/Si-N ratios continued to be linearly related to the Si/N ratio, suggesting[3] that this linear relationship fits over a wide range of deposition parameters.

C14.4   Silicon Oxynitride Films as a Selective Diffusion Barrier

Hashimoto et al.[6] report a study of silicon oxynitride with the expectation that the PECVD film will act as a selective diffusion barrier in the thermal impurity process.

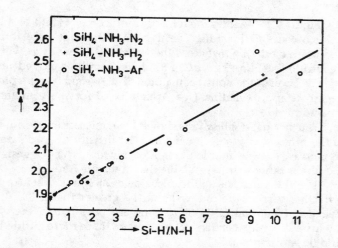

Figure C14-6. Correlation between the refractive index and the Si-H/N-H ratio. (After Claassen[3])

The substrates were (100) n-type GaAs wafers. SiO_xN_y films about 1500 Å thick were formed by PECVD from a mixture of SiH_4, NH_3, and N_2O. The deposition temperature was 300°C, the total gas pressure was 67 Pa and the RF power density was 1.5×10^5 W/m². The flow rates of SiH_4 and NH_3 were 12 and 40 sccm. These parameters were fixed for the entire study. The different compositions were obtained by varying the N_2O flow rates.

Ellipsometry measurements were made using a xenon lamp and a monochromator to give 6358 Å radiation. The angle of incidence was 70°. The composition of the films was determined using Auger electron spectroscopy, taking care to choose the electron beam conditions so as not to alter the composition due to electron beam damage.

The test of the thermal stability of the resulting films was made by thermal annealing in N_2 gas ambient for 6 h at 700°C and looking for microcracks using an optical microscope. The thermal impurity diffusion process was accomplished by means of the closed-tube diffusion technique and was performed for 6 h at 750°C. The performance was characterized by the depth profile of thermally diffused Zn impurities in the films. The depth profile of the Zn was measured by secondary ion mass spectrometry (SIMS).

Figure C14-7 shows the refractive indices, n_f, of SiO_xN_y films as a function of the N_2O flow rate. The higher N_2O flow rates will form a film that is more oxide-like, whereas the lower N_2O flow rates will form a more nitride-like film. Figure C14-8 shows the indices plotted versus the oxygen fraction of the nonsilicon atoms. The dashed line is a plot of the Bruggeman approximation (see Appendix B) for the combination of oxide and nitride.

The thermal stability tests showed that microcracks formed whenever the index was greater than 1.85. The density of cracks decreased with increases of the atomic fraction of oxygen. This suggests that an oxide film is more thermally stable than a nitride film.

The SIMS analysis for the diffusion of Zn shows that in order to be an effective barrier, the index should be greater than 1.80. It is well known[6] that thermal CVD SiO_2 films are porous and therefore not suitable for diffusion barriers. This suggests that a more nitride-like film would be less porous.

Forming a material that is optimum for barrier properties involves a compromise between thermal stability and porosity. The optimum index is between 1.8 and 1.85.

It might be noted in passing that Knolle[7] concludes that for plasma nitrides, plasma oxynitrides, and LPCVD nitrides, the refractive

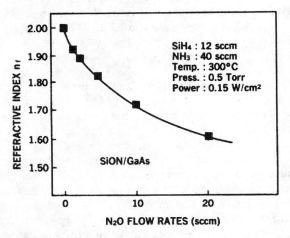

Figure C14-7. The refractive indices of SiO_xN_y films as a function of the N_2O flow rates. (After Hashimoto[6])

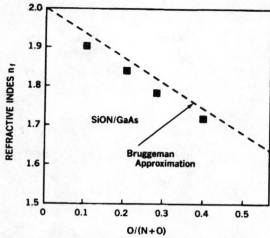

Figure C14-8. The refractive indices of SiO_xN_y films as a function of film composition, $O/(O + N)$. (After Hashimoto[6])

index can be used to estimate the atomic percent silicon in the film. Although their instrumental methods used a prism coupler rather than an ellipsometer, the results are important to this area of investigation. Knolle disagrees with considering a PECVD film to be a combination of oxide and nitride since, in their work, indices were obtained that were greater than 2.0.

C14.5 References

1. G. Lucovsky, M. H. Manitini, J. K. Srivastava, and E. A. Irene, *J. Vac. Sci. Technol. B*, **5**, 530 (1987).
2. W. Kern and D. A. Poutinen, *RCA Rev.*, **31**, 187 (1970).
3. W. A. P. Claassen, W. G. J. N. Valkenburg, F. H. P. M. Habraken, and Y. Tamminga, *J. Electrochem. Soc.*, **130**, 2419 (1983).
4. W. A. Lanford and M. J. Rand, *J. Appl. Phys.*, **49**, 2472 (1978).
5. R. Chow, W. A. Lanford, W. Ke-Ming, and R. S. Rosler, *J. Appl. Phys.*, **53**, 5630 (1982).
6. A. Hashimoto, M. Kobayashi, T. Kamijoh, H. Takano, and M. Sakuta, *J. Electrochem. Soc.*, **133**, 1464 (1986).
7. W. R. Knolle, *Thin Solid Films*, **168**, 123 (1989).

Appendices

APPENDIX A
Del/Psi Trajectory Calculations

A.1 General

Drude[1] shows an apparatus to measure the quantities Del and Psi in a book published in 1901. The fundamental equations of ellipsometry were also understood at that time. To calculate the Del/Psi values for a given structure, one simply applies the Fresnel equations (Equation 14, Chapter 1)

$$r_{12}^p = \frac{\tilde{N}_2 \cos\phi_1 - \tilde{N}_1 \cos\phi_2}{\tilde{N}_2 \cos\phi_1 + \tilde{N}_1 \cos\phi_2} \qquad r_{12}^s = \frac{\tilde{N}_1 \cos\phi_1 - \tilde{N}_2 \cos\phi_2}{\tilde{N}_1 \cos\phi_1 + \tilde{N}_2 \cos\phi_2} \qquad (1)$$

and the Drude equation (Equation 19, Chapter 1)

$$R^p = \frac{r_{12}^p + r_{23}^p \exp(-j\,2\beta)}{1 + r_{12}^p\, r_{23}^p \exp(-j\,2\beta)} \qquad R^s = \frac{r_{12}^s + r_{23}^s \exp(-j\,2\beta)}{1 + r_{12}^s\, r_{23}^s \exp(-j\,2\beta)} \qquad (2)$$

where

$$\beta = 2\pi \left(\frac{d}{\lambda}\right) \tilde{N}_2 \cos\phi_2 \ . \qquad (3)$$

The quantity, rho, is calculated (Equation 24, Chapter 1)

$$\rho = \frac{R^p}{R^s} \qquad (4)$$

233

and Del and Psi are obtained (Equation 25, Chapter 1)

$$\rho = \tan \Psi \, e^{j\Delta} \qquad \text{or} \qquad \tan \Psi \, e^{j\Delta} = \frac{R^p}{R^s} . \qquad (5)$$

Although it was possible to measure Δ and Ψ and make this calculation over 90 years ago, the complex numbers made even a single calculation rather cumbersome. To show how the Del/Psi point moves as a film grows, many calculations are required. This is probably why the development of the mainframe computer spurred the use of ellipsometry in the 1960s. McCrackin[2] published a National Bureau of Standards note that was essentially a FORTRAN listing of a program that would make many of the needed calculations.

Computer programming style had not been developed to today's standards and it is sometimes difficult to find one's way around in the original listing. In addition, the original program was written to be typed on punched cards and run as a batch program.

The original program made many different calculations. This author found that the most useful part of that program was a calculation that McCrackin called "CTABLE," which accepted the optical parameters of the surface and film and then calculated the corresponding values of Del and Psi for a successively thicker film, i.e., the "trajectory" of the Del/Psi point as a function of film thickness.

In the 1990s, desktop computers can make these calculations as readily as mainframe computers did in the 1960s. Correspondingly, Tables A-1 and A-2 are listings of FORTRAN programs which give the Del/Psi trajectories for various thicknesses of the top film for one or three films, respectively. For two films, simply give the middle film a thickness of zero. Although the author uses one particular brand of microcomputer, the listing can be compiled and used on any microcomputer.

A.2   Program OneFilm

The structure is shown at the top of the program. Note that in this listing, the ambient is considered medium "0" rather than medium "1" as is the case in the rest of this text. The first part of the program is the declarations of the variables. Note that we are ignoring the old traditional FORTRAN names for integers, reals, etc., and simply declaring all of the variables in the same manner as some of the more structured languages. Several functions are defined so that the angles can be given in degrees. Next, startup values are given to the variables. Since the author uses only one wavelength and one angle of incidence, the modified program was not written to allow these to be changed.

The program displays the current values of the indices on the screen and allows for the possibility of changing them. Rather than inputting the values of the substrate, the program asks for the film-free Del/Psi values and then calculates the substrate index. After the film thickness values are accepted, the cosine values are calculated from Snell's law and all of the Fresnel coefficients are calculated.

A text file for storing the results is then opened and header information is written to the file. In a "DO Loop," the total reflection coefficients are then calculated for the top surface, rho is determined, and Del and Psi are calculated and written to the file. The information is in tabular format so that it can be moved to a graphing program with "cut & paste."

A.3   Program ThreeFilms

Much of this program is similar to the previous one. Note how the program starts by using the Fresnel coefficients for the bottom two interfaces to determine a Total Reflection coefficient. This in turn is used with the Fresnel coefficient of the next higher interface to determine the Total Reflection coefficient for that next higher interface, etc. The Total Reflection coefficient for the top interface is then used to determine rho. Note that when multiple films are present, the thicknesses of the underlying films are fixed inputs and are not iterated.

A.4   References

1. P. Drude, "The Theory of Optics," Longmans, Green, and Co., New York (1901), p. 258.
2. F. L. McCrackin, *Nat. Bur. Stand.Tech. Note 479* (1969).

TABLE A-1. FORTRAN Listing for calculating the Del/Psi trajectory for a single film on a substrate.

```
            PROGRAM ONEFILM
!               ambient              N0
!                                    ─────────────── rp10, crp10
!               Film        D1    ///// N1 ////
!                                    ─────────────── rp21
!               Substrate         ///// N2 ////

!  Calculations for Ellipsometer with one film on a substrate

!  Indices and Thicknesses
        REAL N0
        COMPLEX N1, N2
        REAL D1, DL, DI, DU

!  Sines and Cosines
        REAL S0, C0
        COMPLEX C1, C2

!  Fresnel Coefficients
        COMPLEX rp10, rn10, rp21, rn21

!  Total Reflection Coefficients (Drude)
        COMPLEX  crp10, crn10

!  Ellipsometry Variables
        COMPLEX RHO
        REAL DEL, DELS, PSI, PSIS

!  Misc. working variables
        CHARACTER CHOICE
        CHARACTER FILENAME*8
        CHARACTER*1 TAB
        REAL ANS11, ANS12
        COMPLEX T1,T2

!  Define Functions
!    SIN, COS, ATAN, TAN for angles in degrees
        SIND(X) = SIN(X/57.29578)
        COSD(X) = COS(X/57.29578)
        ATAND(X) = ATAN(X)*57.29578
        TAND(X)=SIND(X)/COSD(X)
```

(continued)

```
        TAB = CHAR(9)

! Give Standard Values to Variables
        N0=1.
        N1=(1., 0)
        N2=(1., 0)
        AI=70; WL=6328
        DELS=0.0 ; PSIS=0.0
        S0 = SIND(AI) ; C0 = COSD(AI)

! The COMMAND Menu
  10    CALL CLEARSCREEN
  14    FORMAT('{',F5.3,' 'F6.3,'}')
  16    FORMAT(F5.0,' Angstroms')
        type 'Index of the FILM is ';WRITE(9,14) N1
        write(9,*)
        type 'Index of the substrate is ';WRITE(9,14) N2
        WRITE (9, 29) DELS, PSIS
  29    FORMAT(' For the Substrate, DEL = ',F6.2,' PSI = ',F6.2)

        write(9,*)
        WRITE(9,*) '  Do you want to :'
        write(9,*)
        WRITE(9,*) '      (1) - Calculate n & k for the Substrate?'
        write(9,*)
        WRITE(9,*) '      (2) - Input n & k for the Film?'
        write(9,*)
        WRITE(9,*) '      (3) - Calculate TABLE?'
        write(9,*)
        WRITE(9,*) '      (4) - Exit Program?'
        write(9,*)
        TYPE 'CHOOSE Which One :' ; ACCEPT CHOICE

        SELECT CASE (CHOICE)
         CASE("1")
               write(9,*)
               TYPE 'Type in values for DEL & PSI :'
               READ(9,*) DEL, PSI
               DELS = DEL ; PSIS = PSI

               RHO = TAND(PSI)*CEXP((0.0, 1.0)*DEL/57.29578)
               N2=N0*S0/C0*CSQRT(1.0-4*RHO*S0**2/(RHO+1.0)**2)

         CASE("2")
               write(9,*)
               TYPE 'Type n & k for the top film :'
               READ (9,*) ANS11, ANS12
               N1 = CMPLX(ANS11, -ANS12)
```
(continued)

```
        CASE("3")
              GOTO 200
100           CONTINUE

        CASE("4")
              WRITE(9,*)
              TYPE "End Program?  <Y or N>:"; ACCEPT CHOICE
              IF (CHOICE.EQ."Y".OR.CHOICE.EQ."y") THEN
                    STOP
              END IF
        CASE DEFAULT
              WRITE(9,*) 'You must choose 1, 2, 3, or 4!'
              TYPE "Hit <RETURN> to continue."; PAUSE
        END SELECT
        GOTO 10
!=====================================================

! Calculating TABLE

200     CALL CLEARSCREEN

!             Inputing Values
        TYPE 'Type the LOWER Limit of film thickness :';ACCEPT DL
        TYPE 'Type the thickness INCREMENT :';ACCEPT DI
        TYPE 'Type the UPPER LIMIT of film thickness :';ACCEPT DU
        write(9,*)
        WRITE(9,210) DL, DU, DI
210     FORMAT(' From ',F7.0,' to ',F7.0,' in steps of ',F5.1)
        write(9,*)
        TYPE(9,*) 'Is this correct?  <Y or N> :';ACCEPT CHOICE
        IF(CHOICE.EQ."Y".OR.CHOICE.EQ."y") THEN
              GOTO 220
          ELSE
              GOTO 200
        END IF
220     DU = DU*1.000001

! Calculate Cosines
        C1 = CSQRT( 1. - (S0*N0/N1)**2 )
        C2 = CSQRT( 1. - (S0*N0/N2)**2 )

! Calculate Fresnel Coefficients

        rp10 = ( N1*C0 - N0*C1 )/( N1*C0 + N0*C1 )
        rn10 = ( N0*C0 - N1*C1 )/( N0*C0 + N1*C1 )
        rp21 = ( N2*C1 - N1*C2 )/( N2*C1 + N1*C2 )
        rn21 = ( N1*C1 - N2*C2 )/( N1*C1 + N2*C2 )
```

                                                        (continued)

```
!                   Opening and Writing the Calc. to File
            write(9,*)
            TYPE 'Filename to store Data? :';ACCEPT FILENAME

            OPEN(44,FILE=FILENAME,STATUS='NEW')


!               Type in header information
            WRITE(44,*) ' ELLIPSOMETER CALCULATIONS'
            write(44,*)
            WRITE(44,310) AI, WL
310         FORMAT(' Angle of Incidence = ',F5.1,' Wavelength = ',F6.0)
            write(44,*)
            WRITE(44,312) N0
312         FORMAT(' Index of Medium = ',F6.3)
            WRITE(44,316) N1
316         FORMAT(' Index of the FILM is {',F5.3,' ',F6.3,'}')
            write(44,326) N2
326         FORMAT(' Index of the SUBSTRATE is {',F5.3,' ',F6.3,'}')
            write(44,*)
            WRITE(44,*)' TH',TAB,TAB,' PSI',TAB,TAB,' DEL'
            write(44,*)

!               Do calculations and write them to the file.

            DO (D1=DL, DU, DI)

              T1 = CEXP((0.0, -12.56637)*N1*C1*D1/WL)
              crp10 =( rp10 + rp21*T1 )/( 1 + rp10*rp21*T1 )
              crn10 =( rn10 + rn21*T1 )/( 1 + rn10*rn21*T1 )

              RHO = crp10/crn10
              DEL = ATAN2(AIMAG(RHO), REAL(RHO))*57.29578
              IF(REAL(RHO).LT.0..AND.ABS(AIMAG(RHO)).LT.1E-6) DEL = 180
              PSI=ATAND(CABS(RHO))
              IF(DEL.LT.0.0) DEL = DEL + 360
              WRITE(44,*) D1,TAB, PSI,TAB, DEL
            REPEAT

            CLOSE(44)

            GOTO 100
            END
```

(continued)

```
!================================================

        SUBROUTINE CLEARSCREEN
        DO (30 TIMES)
                WRITE (9,*)
        REPEAT

        END
```

TABLE A-2. FORTRAN Listing for calculating the Del/Psi trajectory for three films on a substrate.

```
        PROGRAM THREEFILMS
```

```
!               ambient                    N0
!                                        ——————————— rp10, crp10
!               Top film        D1      ///// N1 ////
!                                        ——————————— rp21, crp21
!               Middle film     D2      \\\\ N2 \\\\
!                                        ——————————— rp32, crp32
!               Bottom film     D3      ///// N3 ////
!                                        ——————————— rp43
!                                        \\\\ N4 \\\\
```

```
! Calculations for Ellipsometer with three films on substrate

! Indices and Thicknesses
        REAL N0
        COMPLEX N1, N2, N3, N4
        REAL D1, D2, D3, DL, DI, DU
```

(continued)

```
!  Sines and Cosines
        REAL S0, C0
        COMPLEX C1, C2, C3, C4

!  Fresnel Coefficients
        COMPLEX rp10, rn10, rp21, rn21, rp32, rn32, rp43, rn43

!  Total Reflection Coefficients (Drude)
        COMPLEX crp32, crn32, crp21, crn21, crp10, crn10

!  Ellipsometery Variables
        COMPLEX RHO
        REAL DEL, DELS, PSI, PSIS

!  Misc. working variables
        CHARACTER CHOICE
        CHARACTER FILENAME*8
        CHARACTER*1 TAB
        REAL ANS11, ANS12, ANS21, ANS22, ANS31, ANS32
        COMPLEX T1,T2,T3

!  Define Functions
!    SIN, COS, ATAN, TAN for angles in degrees
        SIND(X) = SIN(X/57.29578)
        COSD(X) = COS(X/57.29578)
        ATAND(X) = ATAN(X)*57.29578
        TAND(X)=SIND(X)/COSD(X)

        TAB = CHAR(9)

!  Give Standard Values to Variables
        N0=1.
        N1=(1., 0)
        N2=(1., 0)
        N3=(1., 0)
        N4=(1., 0)
        D2=0; D3=0
        AI=70;  WL=6328
        DELS=0.0 ; PSIS=0.0
        S0 = SIND(AI) ; C0 = COSD(AI)

!  The COMMAND Menu
    10    CALL CLEARSCREEN
    14    FORMAT('{',F5.3,' 'F6.3,'}')
    16    FORMAT(F5.0,' Angstroms')
          type 'Index of the TOP film is ';WRITE(9,14) N1
```

(continued)

```
        write(9,*)
        type 'Index of the MIDDLE film is ';WRITE(9,14) N2
        type '  Thickness of the MIDDLE film is ';WRITE(9,16) D2
        write(9,*)
        type 'Index of the BOTTOM film is ';WRITE(9,14) N3
        type '  Thickness of the BOTTOM film is ';WRITE(9,16) D3
        write(9,*)
        type 'Index of the substrate is ';WRITE(9,14) N4
        WRITE (9, 29) DELS, PSIS
29      FORMAT('  For the Substrate, DEL = ',F6.2,'  PSI = ',F6.2)


        write(9,*)
        WRITE(9,*) '  Do you want to :'
        write(9,*)
        WRITE(9,*) '        (1) - Calculate n & k for the Substrate?'
        write(9,*)
        WRITE(9,*) '        (2) - Input Bottom Film Parameters?'
        write(9,*)
        WRITE(9,*) '        (3) - Input Middle Film Parameters?'
        write(9,*)
        WRITE(9,*) '        (4) - Input n & k for the Top Film?'
        write(9,*)
        WRITE(9,*) '        (5) - Calculate TABLE?'
        write(9,*)
        WRITE(9,*) '        (6) - Exit Program?'
        write(9,*)
        TYPE 'CHOOSE Which One :' ; ACCEPT CHOICE

        SELECT CASE (CHOICE)
         CASE("1")
                write(9,*)
                TYPE 'Type in values for DEL & PSI :'
                READ(9,*) DEL, PSI
                DELS = DEL ; PSIS = PSI

                RHO = TAND(PSI)*CEXP((0.0, 1.0)*DEL/57.29578)
                N4=N0*S0/C0*CSQRT(1.0-4*RHO*S0**2/(RHO+1.0)**2)

         CASE("2")
                write(9,*)
                TYPE 'Type n & k for the bottom film :'
                READ (9,*) ANS31, ANS32
                N3 = CMPLX(ANS31, -ANS32)
                TYPE 'Type the THICKNESS of the bottom film :'
                READ (9,*) D3
```

(continued)

```
        CASE("3")
            write(9,*)
            TYPE 'Type n & k for the middle film :'
            READ (9,*) ANS21, ANS22
            N2 = CMPLX(ANS21, -ANS22)
            TYPE 'Type the THICKNESS of the middle film :'
            READ (9,*) D2

        CASE("4")
            write(9,*)
            TYPE 'Type n & k for the top film :'
            READ (9,*) ANS11, ANS12
            N1 = CMPLX(ANS11, -ANS12)

        CASE("5")
            GOTO 200
100         CONTINUE

        CASE("6")
            WRITE(9,*)
            TYPE "End Program? <Y or N>:"; ACCEPT CHOICE
            IF (CHOICE.EQ."Y".OR.CHOICE.EQ."y") THEN
                STOP
            END IF
        CASE DEFAULT
            WRITE(9,*) 'You must choose 1, 2, 3, 4, 5,or 6!'
            TYPE "Hit <RETURN> to continue."; PAUSE
        END SELECT
        GOTO 10

!=====================================================

! Calculating TABLE

200     CALL CLEARSCREEN

!           Inputing Values
        TYPE 'Type the LOWER Limit of film thickness :';ACCEPT DL
        TYPE 'Type the thickness INCREMENT :';ACCEPT DI
        TYPE 'Type the UPPER LIMIT of film thickness :';ACCEPT DU

        write(9,*)
        WRITE(9,210) DL, DU, DI
210     FORMAT(' From ',F7.0,' to ',F7.0,' in steps of ',F5.1)
        write(9,*)
        TYPE(9,*) 'Is this correct?  <Y or N> :';ACCEPT CHOICE
```

(continued)

```
          IF(CHOICE.EQ."Y".OR.CHOICE.EQ."y") THEN
               GOTO 220
          ELSE
               GOTO 200
          END IF
220      DU = DU*1.000001

!  Calculate Cosines
          C1 = CSQRT( 1. - (S0*N0/N1)**2 )
          C2 = CSQRT( 1. - (S0*N0/N2)**2 )
          C3 = CSQRT( 1. - (S0*N0/N3)**2 )
          C4 = CSQRT( 1. - (S0*N0/N4)**2 )

!  Calculate Fresnel Coefficients

          rp10 = ( N1*C0 - N0*C1 )/( N1*C0 + N0*C1 )
          rn10 = ( N0*C0 - N1*C1 )/( N0*C0 + N1*C1 )
          rp21 = ( N2*C1 - N1*C2 )/( N2*C1 + N1*C2 )
          rn21 = ( N1*C1 - N2*C2 )/( N1*C1 + N2*C2 )
          rp32 = ( N3*C2 - N2*C3 )/( N3*C2 + N2*C3 )
          rn32 = ( N2*C2 - N3*C3 )/( N2*C2 + N3*C3 )
          rp43 = ( N4*C3 - N3*C4 )/( N4*C3 + N3*C4 )
          rn43 = ( N3*C3 - N4*C4 )/( N3*C3 + N4*C4 )

!  Calculate Total Reflection Coefficients for all but top

          T3 = CEXP((0.0, -12.56637)*N3*C3*D3/WL)
          crp32 =( rp32 + rp43*T3 )/( 1 + rp32*rp43*T3 )
          crn32 =( rn32 + rn43*T3 )/( 1 + rn32*rn43*T3 )

          T2 = CEXP((0.0, -12.56637)*N2*C2*D2/WL)
          crp21 =( rp21 + crp32*T2 )/( 1 + rp21*crp32*T2 )
          crn21 =( rn21 + crn32*T2 )/( 1 + rn21*crn32*T2 )


!                  Opening and Writing the Calc. to File
          write(9,*)
          TYPE 'Filename to store Data? :';ACCEPT FILENAME

          OPEN(44,FILE=FILENAME,STATUS='NEW')


!          Type in header information
          WRITE(44,*) ' ELLIPSOMETER CALCULATIONS'
          write(44,*)
          WRITE(44,310) AI, WL
```

(continued)

```
310   FORMAT(' Angle of Incidence = ',F5.1,'  Wavelength = ',F6.0)
      write(44,*)
      WRITE(44,312) N0
312   FORMAT(' Index of Medium = ',F6.3)
      WRITE(44,316) N1
316   FORMAT('  Index of the TOP Film is {',F5.3,' ',F6.3,'}')
      WRITE(44,318) N2
318   FORMAT('  Index of the MIDDLE Film is {',F5.3,' ',F6.3,'}')
      write(44,320) D2
320   FORMAT(' Thickness of the middle film is ',F5.0,' angstroms')
      WRITE(44,322) N3
322   FORMAT('  Index of the BOTTOM Film is {',F5.3,' ',F6.3,'}')
      write(44,324) D3
324   FORMAT(' Thickness of the bottom film is ',F5.0,' angstroms')
      write(44,326) N4
326   FORMAT('  Index of the SUBSTRATE is {',F5.3,' ',F6.3,'}')
      write(44,*)
      WRITE(44,*)' TH',TAB,TAB,' PSI',TAB,TAB,' DEL'
      write(44,*)

      Do calculations and write them to the file.

      DO (D1=DL, DU, DI)
      T1 = CEXP((0.0, -12.56637)*N1*C1*D1/WL)
      crp10 =( rp10 + crp21*T1 )/( 1 + rp10*crp21*T1 )
      crn10 =( rn10 + crn21*T1 )/( 1 + rn10*crn21*T1 )

      RHO = crp10/crn10
      DEL = ATAN2(AIMAG(RHO), REAL(RHO))*57.29578
      IF(REAL(RHO).LT.0..AND.ABS(AIMAG(RHO)).LT.1E-6) DEL = 180
      PSI=ATAND(CABS(RHO))
      IF(DEL.LT.0.0) DEL = DEL + 360
      WRITE(44,*) D1,TAB, PSI,TAB, DEL
      REPEAT

      CLOSE(44)

      GOTO 100
      END

==================================================

      SUBROUTINE CLEARSCREEN
      DO (30 TIMES)
            WRITE (9,*)
      REPEAT

      END
```

APPENDIX B
Effective Medium Considerations

B.1   General

Several situations arise when a layer is not homogeneous, but consists of a combination of two or more different materials. Minor inclusions in a matrix is one example. Another example is a rough surface where the mean height and correlation length of the irregularities are both less than the wavelength of light. The combination of the material and the voids can be considered to make up an "effective layer" as suggested in Figure B-1. It is the purpose of this appendix to show how to calculate the complex index of refraction for the "effective medium" in cases such as these.

Numerous works have considered this matter. In 1979, D. E. Aspnes and co-workers wrote a rather comprehensive review[1] of most of the previous works and most of what follows in this appendix summarizes some of this work.

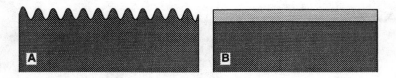

Figure B-1. (A) A rough surface and (B) the equivalent effective medium on a substrate of the original material.

B.2 . Theory

The Lorentz-Lorenz (LL), Maxwell Garnett (MG), and Bruggeman effective medium approximation (EMA) models are simple effective medium theories[2,3] that represent heterogeneous mixtures. They all have the same form

$$\frac{\tilde{N}_e^2 - \tilde{N}_h^2}{\tilde{N}_e^2 + 2\tilde{N}_h^2} = f_1 \frac{\tilde{N}_1^2 - \tilde{N}_h^2}{\tilde{N}_1^2 + 2\tilde{N}_h^2} + f_2 \frac{\tilde{N}_2^2 - \tilde{N}_h^2}{\tilde{N}_2^2 + 2\tilde{N}_h^2} + \dots \tag{1}$$

where $\tilde{N}_e$, $\tilde{N}_h$, $\tilde{N}_1$, $\tilde{N}_2$ are the complex indices for the effective medium, the host medium and inclusions of types 1, 2, etc. Note that we have not yet defined the "host" medium. The f_1, f_2, etc., represent the volume fractions of inclusions 1, 2, etc. We shall see that the primary difference in the LL, MG, and EMA models is choice of the "host" medium.

The underlying assumptions are spherical inclusion geometry and dipole interactions.[2,3] Although neither assumption is usually satisfied rigorously, these are standard first approximations that usually give good results.

The LL approximation was developed to describe point polarizable entities embedded in vacuum. In this case, $\tilde{N}_h = 1$. The equation becomes

$$\frac{\tilde{N}_e^2 - 1}{\tilde{N}_e^2 + 2} = f_1 \frac{\tilde{N}_1^2 - 1}{\tilde{N}_1^2 + 2} + f_2 \frac{\tilde{N}_2^2 - 1}{\tilde{N}_2^2 + 2} \tag{2}$$

For the roughness involved in island formation film growth, this is a reasonable approximation up to a volume fraction of about 20%.

The MG approximation[4] corresponds to inclusions in a host background (other than vacuum) and the quantities in Equation 1 have their obvious interpretations. In the case of a single inclusion in a single host, the equation becomes

$$\frac{\tilde{N}_e^2 - \tilde{N}_h^2}{\tilde{N}_e^2 + 2\tilde{N}_h^2} = f_1 \frac{\tilde{N}_1^2 - \tilde{N}_h^2}{\tilde{N}_1^2 + 2\tilde{N}_h^2} \tag{3}$$

This degenerates to the LL form if the host happens to be vacuum or air. Again, this is not a bad approximation if the inclusions make up a small fraction of the total volume.

Aspnes *et al.*[1] point out how this theory breaks down when using this model in rough surface applications or in other applications where there is about as much inclusion as there is host. For a single type of inclusion, one calculates different values of $\tilde{N}_e$ if the roles of host and inclusion are interchanged, even if the respective volume fractions stay the same.[3]

To deal with this matter, Bruggeman[5] suggested making the "host" the effective medium itself, i.e., making $\tilde{N}_h = \tilde{N}_e$. With this formulation, Equation 1 becomes

$$f_1 \frac{\tilde{N}_1^2 - \tilde{N}_e^2}{\tilde{N}_1^2 + 2\tilde{N}_e^2} + f_2 \frac{\tilde{N}_2^2 - \tilde{N}_e^2}{\tilde{N}_2^2 + 2\tilde{N}_e^2} = 0 \ . \tag{4}$$

This is the effective medium approximation (EMA) equation for two materials. It can easily be extended to more materials in the obvious way.

B.3 Examples

To show how the EMA and the MG approximation compare, let us consider the case of a mixture of silicon with $\tilde{N} = 3.85 - 0.02j$ and air with $\tilde{N} = 1.0 - 0.0j$. The MG approximation degenerates to the LL approximation for air. Let us calculate the value of $\tilde{N}_e$ as we vary the amount of silicon from zero to 1.0. For the MG (or LL) approximation, silicon will be inclusion 1 and air will be the host. For the EMA, silicon will be material 1 and air will material 2. The results are plotted in Figure B-2. The results are similar up to a silicon fraction of about 0.2, after which they begin to deviate significantly. The strong asymmetry of the MG model is due to the fact that treating air as the host when the silicon fraction nears 1.0 is clearly stretching the imagination. Aspnes[1] makes the point that the EMA model makes more sense near the center of these curves and he goes on to use spectroscopic ellipsometry measurements on polycrystalline silicon to illustrate that this is the case.

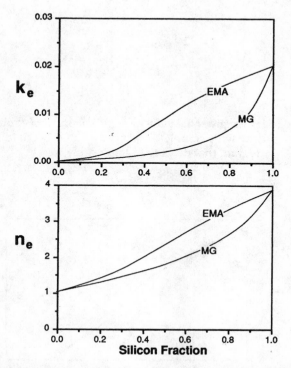

Figure B-2. Values of the effective index of refraction and extinction coefficient for a mixture of silicon and air calculated with the effective medium approximation (EMA) and the Maxwell Garnett (MG) approximation.

Although somewhat messy, simple algebra can be used to convert Equation 4 to

$$\tilde{N}_1^2\tilde{N}_2^2 + \left(\tilde{N}_1^2(2f_1 - f_2) + \tilde{N}_2^2(2f_2 - f_1)\right)\tilde{N}_e^2 - 2\tilde{N}_e^2\tilde{N}_e^2 = 0 \ . \quad (5)$$

In this form, the quadratic equation can be used to solve for $\tilde{N}_e^2$, i.e.,

$$\tilde{N}_e^2 = \frac{-B \pm \sqrt{B^2 - 4AC}}{2A} \quad (6)$$

where

$$A = -2$$

$$B = \tilde{N}_1^2(2f_1 - f_2) + \tilde{N}_2^2(2f_2 - f_1)$$

$$C = \tilde{N}_1^2 \tilde{N}_2^2 \tag{7}$$

A square root then gives the value of $\tilde{N}_e$. One must, of course, have the facilities to deal with complex numbers. Table B-1 is a listing of a FORTRAN program that deals with this readily.

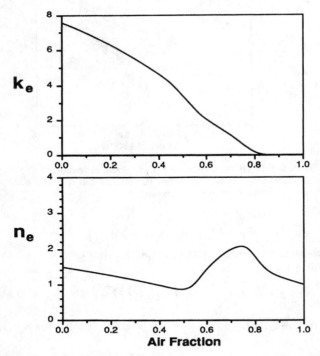

Figure B-3. Values of the effective index of refraction and extinction coefficient for a mixture of aluminum and air calculated with the effective medium approximation. For aluminum, we use[6] $\tilde{N} = 1.5 - 7.6\,j$.

Because of the "±" in the quadratic equation, we calculate two values of $\tilde{N}_e{}^2$ and subsequently of $\tilde{N}_e$. One will be meaningful (as we would expect an index to be, i.e., first term positive, second term negative) and the other will be meaningless.

As a second example, we calculate the effective index using a material with a high value of k. We choose a mixture of aluminum and air. For aluminum, we take[6] the index to be $\tilde{N} = 1.5 - 7.6 \, j$, and for air, of course, we take $\tilde{N} = 1.0 - 0.0 \, j$. The resulting graph is shown as Figure B-3. We see that for k, the curve decreases monotonically from the high value for Al to zero for air. The curve for n moves smoothly, but not monotonically from that of Al to that of air.

B.4   References

1. D. E. Aspnes, J. B. Theeten, and F. Hottier, *Phys. Rev. B*, **20**, 3292 (1979).
2. C. G. Grandqvist and O. Hunderi, *Phys. Rev. B*, **16**, 3513 (1977).
3. R. Landauer, in "Proceedings of the First Conference on Electrical Transport and Optical Properties of Inhomogeneous Media," edited by J. C. Garland and D. B. Tanner, AIP Conf. Proc. No. 40, AIP, New York (1978).
4. J. C. Maxwell Garnett, *Philos. Trans. R. Soc. London*, **203**, 385 (1904); **A205**, 237 (1906).
5. D. A. G. Bruggeman, *Ann. Phys. (Leip.)*, **24**, 636 (1935).
6. E. D. Palik, "Handbook of Optical Constants of Solids," Academic Press, New York (1985).

TABLE B-1. FORTRAN Listing for calculating Index of Refraction using the Bruggeman Effective Medium Approximation.

```
PROGRAM EffMedium

! Declare Variables and Assign Constants
      COMPLEX N1, N2, Nsq_pos, Nsq_neg, Ne_pos, Ne_neg
      COMPLEX A, B, C
      REAL f1, f2
      CHARACTER ANS

      ANS='Y'
      A = CMPLX(-2.0, 0)
```

(continued)

```
! Input Data
        DO (5 TIMES); WRITE(9,*); REPEAT
        WHILE (ANS .EQ. 'Y' .OR. ANS .EQ. 'y')

                WRITE(9,*)
                WRITE(9,*) '   Type in new values '
                TYPE 'For N1 :'
                READ(9,*) N1
                TYPE 'For f1 :'
                READ(9,*) f1
                f2 = 1.0 - f1
                TYPE 'For N2 :'
                READ(9,*) N2

! Do Calculation
                B = N1*N1*(2*f1-f2)+N2*N2*(2*f2-f1)
                C = N1*N1*N2*N2

                Nsq_pos = (-B + CSQRT(B*B-4*A*C))/(2*A)
                Nsq_neg = (-B - CSQRT(B*B-4*A*C))/(2*A)

                Ne_pos = CSQRT(Nsq_pos)
                Ne_neg = CSQRT(Nsq_neg)

! Print it on Screen
                WRITE(9,*)
                WRITE(9,*) 'f1 = ', f1
                WRITE(9,*) 'N1 = ', N1
                WRITE(9,*)
                WRITE(9,*) 'f2 = ', f2
                WRITE(9,*) 'N2 = ', N2
                WRITE(9,*)
                WRITE(9,*) 'Ne_pos = ', Ne_pos
                WRITE(9,*) 'Ne_neg = ', Ne_neg
                WRITE(9,*)

                TYPE 'Do Again ?? < y or n > :'
                READ(9,*) ANS
        REPEAT

        STOP
        END
```

APPENDIX C
Literature Values of Optical Constants of Various Materials

C.1 General

The amount of material in the technical literature concerning ellipsometry and other optical techniques is quite large. Although in several places in this text there are warnings concerning using handbook values for the optical constants, we list here various values from the reviewed literature. The purpose of this listing is to give the reader a ballpark value for comparisons only. The list is not exhaustive, but is primarily for those materials encountered in the literature during the production of this book. Mention should be made of the two volumes[1,2] by E. D. Palik. Numerous values were obtained from these handbooks. The values listed herein are for a wavelength of 6328 Å. In some cases the values listed on either side of 6328 Å were interpolated to give an estimate of the value at 6328 Å.

C.2 Values

Material	n	k	Ref.
Ag	0.13	3.99	1
Ag	0.13	3.49	6
Al	1.49	7.34	1
Al	1.37	7.62	1
Al_2O_3	1.77	0	2
AlAs	3.11	0.00	2
AlON	1.79	0	2
Au	0.18	3.10	1
Be	3.46	3.19	2
Bi	1.8	4.0	7
Bi_2O_3	2.6	0	7

Material	n	k	Ref.
CaF$_2$	1.43	0	2
CdS	2.38	0.52	2
CdSe	2.74	0.27	2
Co	2.21	4.17	2
Co	1.96	4.16	2
Cr	3.58	4.36	2
CsI	1.78	0	2
Cu	0.25	3.41	1
Cu$_2$O	2.94	0.11	2
CuO	2.70	0.43	2
Diamond	2.41	0	1
Fe	2.43	3.3	9
GaAs	3.86	0.20	1
H$_2$O	1.33	0	2
Hg	1.99	5.24	2
InAs	3.96	0.61	1
InP	3.54	0.31	1
InP	3.52	0.3	17
InP	3.47	0.45	18
InSb	4.25	1.80	1
Ir	2.53	4.61	1
Ir	2.52	5.09	12
K	0.05	1.75	2
KBr	1.56	0	2
KCl	1.49	0	1
Li	0.22	2.93	2
LiF	1.40	0	1
MgO	1.735	0	2
Mo	3.70	3.54	1
Mo	4.01	3.86	16
Na	0.05	2.64	2
NaCl	1.542	0	1
NaF	1.325	0	2
Nb	2.83	2.86	2
Nb	3.03	3.61	11
Ni	1.97	3.72	1
Os	3.90	1.66	1
PbS	4.338	1.444	1
PbSe	3.970	3.032	1
PbTe	6.352	4.060	1
Pd	1.77	4.29	2
Pt	2.33	4.15	1
Rh	2.15	5.61	1
a-Si	4.206	0.422	1
Si	3.882	0.019	1
Si$_3$N$_4$	2.021	0	1

Material	n	k	Ref.
SiC	2.632	0.000	1
SiC	2.660	0.000	2
a-SiO$_2$	1.457	0	1
c-SiO$_2$	1.542	0	1
SiO	1.965	0.010	1
SnTe	3.257	5.311	2
Ta	1.72	2.09	2
Ta	2.3	2.6	8
Ta	3.02	2.57	11
Ta oxide	2.22	0	8
Ti	3.23	3.62	15
Ti:W	2.84	3.08	4
Ti:W oxide	2.25	0.12	4
TiN	1.39	1.76	5
TiO$_2$	2.2	0	5
V	3.53	2.95	2
V	3.637	3.334	10
V	3.838	3.56	14
W	3.64	2.91	1
W	4.3	3.1	11
WSi$_{2.2}$	4.25	1.37	3
Xe	1.48	0	6
ZnTe	2.980	0.075	2
Zr	2.21	3.04	13
Zr oxide	2.17	0.033	13

C.3   References

1. "Handbook of Optical Constants of Solids", edited by E. D. Palik, Academic Press, New York (1985).
2. "Handbook of Optical Constants of Solids II," edited by E. D. Palik, Academic Press, New York (1991).
3. H. G. Tompkins, *Thin Solid Films*, **181**, 285 (1989).
4. H. G. Tompkins and S. Lytle, *J. Appl. Phys.*, **64**, 3269 (1988).
5. H. G. Tompkins, *J. Appl. Phys.*, **70**, 3876 (1991)
6. A. Itakura and I. Arakawa, *J. Vac. Sci. Technol. A*, **9**, 1779 (1991).
7. R. Atkinson and E. Curran, *Thin Solid Films*, **128**, 333 (1985).
8. J. D. Leslie and K. Knorr, *J. Electrochem. Soc.*, **121**, 263 (1974).
9. T. Zakroczymski, C.-J. Fan, and Z. Szklarska-Smialowska, *J. Electrochem. Soc.*, **132**, 2862 (1985).

10. J. L. Ord, S. D. Bishop, and D. J. De Smet, *J. Electrochem. Soc.*, **138**, 208 (1991).
11. J. L. Ord, M. A. Hopper, and W. P. Wang, *J. Electrochem. Soc.*, **119**, 439 (1972).
12. J. L. Ord, *J. Electrochem. Soc.*, **129**, 335 (1982).
13. M. A. Hopper, J. A. Wright, and D. J. DeSmet, *J. Electrochem. Soc.*, **124**, 44 (1977).
14. J. C. Clayton and D. J. De Smet, *J. Electrochem. Soc.*, **123**, 1886 (1976).
15. J. L. Ord, D. J. De Smet, and D. J. Beckstead, *J. Electrochem. Soc.*, **136**, 2178 (1989).
16. D. J. De Smet and J. L Ord, *J. Electrochem. Soc.*, **134**, 1734 (1987).
17. X. Liu, E. A. Irene, S. Hattangady, and G. Fountain, *J. Electrochem. Soc.*, **137**, 2319 (1990).
18. R. Scheps, *J. Electrochem. Soc.*, **131**, 540 (1984).

INDEX

A CATALOG OF SELECTED
DOVER BOOKS
IN SCIENCE AND MATHEMATICS

Astronomy

BURNHAM'S CELESTIAL HANDBOOK, Robert Burnham, Jr. Thorough guide to the stars beyond our solar system. Exhaustive treatment. Alphabetical by constellation: Andromeda to Cetus in Vol. 1; Chamaeleon to Orion in Vol. 2; and Pavo to Vulpecula in Vol. 3. Hundreds of illustrations. Index in Vol. 3. 2,000pp. 6⅛ x 9¼.
Vol. I: 0-486-23567-X
Vol. II: 0-486-23568-8
Vol. III: 0-486-23673-0

EXPLORING THE MOON THROUGH BINOCULARS AND SMALL TELESCOPES, Ernest H. Cherrington, Jr. Informative, profusely illustrated guide to locating and identifying craters, rills, seas, mountains, other lunar features. Newly revised and updated with special section of new photos. Over 100 photos and diagrams. 240pp. 8¼ x 11. 0-486-24491-1

THE EXTRATERRESTRIAL LIFE DEBATE, 1750–1900, Michael J. Crowe. First detailed, scholarly study in English of the many ideas that developed from 1750 to 1900 regarding the existence of intelligent extraterrestrial life. Examines ideas of Kant, Herschel, Voltaire, Percival Lowell, many other scientists and thinkers. 16 illustrations. 704pp. 5⅜ x 8½. 0-486-40675-X

THEORIES OF THE WORLD FROM ANTIQUITY TO THE COPERNICAN REVOLUTION, Michael J. Crowe. Newly revised edition of an accessible, enlightening book recreates the change from an earth-centered to a sun-centered conception of the solar system. 242pp. 5⅜ x 8½. 0-486-41444-2

A HISTORY OF ASTRONOMY, A. Pannekoek. Well-balanced, carefully reasoned study covers such topics as Ptolemaic theory, work of Copernicus, Kepler, Newton, Eddington's work on stars, much more. Illustrated. References. 521pp. 5⅜ x 8½.
0-486-65994-1

A COMPLETE MANUAL OF AMATEUR ASTRONOMY: TOOLS AND TECHNIQUES FOR ASTRONOMICAL OBSERVATIONS, P. Clay Sherrod with Thomas L. Koed. Concise, highly readable book discusses: selecting, setting up and maintaining a telescope; amateur studies of the sun; lunar topography and occultations; observations of Mars, Jupiter, Saturn, the minor planets and the stars; an introduction to photoelectric photometry; more. 1981 ed. 124 figures. 25 halftones. 37 tables. 335pp. 6½ x 9¼. 0-486-40675-X

AMATEUR ASTRONOMER'S HANDBOOK, J. B. Sidgwick. Timeless, comprehensive coverage of telescopes, mirrors, lenses, mountings, telescope drives, micrometers, spectroscopes, more. 189 illustrations. 576pp. 5⅝ x 8¼. (Available in U.S. only.)
0-486-24034-7

STARS AND RELATIVITY, Ya. B. Zel'dovich and I. D. Novikov. Vol. 1 of *Relativistic Astrophysics* by famed Russian scientists. General relativity, properties of matter under astrophysical conditions, stars, and stellar systems. Deep physical insights, clear presentation. 1971 edition. References. 544pp. 5⅝ x 8¼. 0-486-69424-0

Chemistry

THE SCEPTICAL CHYMIST: THE CLASSIC 1661 TEXT, Robert Boyle. Boyle defines the term "element," asserting that all natural phenomena can be explained by the motion and organization of primary particles. 1911 ed. viii+232pp. 5⅜ x 8½.
0-486-42825-7

RADIOACTIVE SUBSTANCES, Marie Curie. Here is the celebrated scientist's doctoral thesis, the prelude to her receipt of the 1903 Nobel Prize. Curie discusses establishing atomic character of radioactivity found in compounds of uranium and thorium; extraction from pitchblende of polonium and radium; isolation of pure radium chloride; determination of atomic weight of radium; plus electric, photographic, luminous, heat, color effects of radioactivity. ii+94pp. 5⅜ x 8½.   0-486-42550-9

CHEMICAL MAGIC, Leonard A. Ford. Second Edition, Revised by E. Winston Grundmeier. Over 100 unusual stunts demonstrating cold fire, dust explosions, much more. Text explains scientific principles and stresses safety precautions. 128pp. 5⅜ x 8½.   0-486-67628-5

THE DEVELOPMENT OF MODERN CHEMISTRY, Aaron J. Ihde. Authoritative history of chemistry from ancient Greek theory to 20th-century innovation. Covers major chemists and their discoveries. 209 illustrations. 14 tables. Bibliographies. Indices. Appendices. 851pp. 5⅜ x 8½.   0-486-64235-6

CATALYSIS IN CHEMISTRY AND ENZYMOLOGY, William P. Jencks. Exceptionally clear coverage of mechanisms for catalysis, forces in aqueous solution, carbonyl- and acyl-group reactions, practical kinetics, more. 864pp. 5⅜ x 8½.
0-486-65460-5

ELEMENTS OF CHEMISTRY, Antoine Lavoisier. Monumental classic by founder of modern chemistry in remarkable reprint of rare 1790 Kerr translation. A must for every student of chemistry or the history of science. 539pp. 5⅜ x 8½. 0-486-64624-6

THE HISTORICAL BACKGROUND OF CHEMISTRY, Henry M. Leicester. Evolution of ideas, not individual biography. Concentrates on formulation of a coherent set of chemical laws. 260pp. 5⅜ x 8½.   0-486-61053-5

A SHORT HISTORY OF CHEMISTRY, J. R. Partington. Classic exposition explores origins of chemistry, alchemy, early medical chemistry, nature of atmosphere, theory of valency, laws and structure of atomic theory, much more. 428pp. 5⅜ x 8½. (Available in U.S. only.)   0-486-65977-1

GENERAL CHEMISTRY, Linus Pauling. Revised 3rd edition of classic first-year text by Nobel laureate. Atomic and molecular structure, quantum mechanics, statistical mechanics, thermodynamics correlated with descriptive chemistry. Problems. 992pp. 5⅜ x 8½.   0-486-65622-5

FROM ALCHEMY TO CHEMISTRY, John Read. Broad, humanistic treatment focuses on great figures of chemistry and ideas that revolutionized the science. 50 illustrations. 240pp. 5⅜ x 8½.   0-486-28690-8

Engineering

DE RE METALLICA, Georgius Agricola. The famous Hoover translation of greatest treatise on technological chemistry, engineering, geology, mining of early modern times (1556). All 289 original woodcuts. 638pp. 6¾ x 11. 0-486-60006-8

FUNDAMENTALS OF ASTRODYNAMICS, Roger Bate et al. Modern approach developed by U.S. Air Force Academy. Designed as a first course. Problems, exercises. Numerous illustrations. 455pp. 5⅜ x 8½. 0-486-60061-0

DYNAMICS OF FLUIDS IN POROUS MEDIA, Jacob Bear. For advanced students of ground water hydrology, soil mechanics and physics, drainage and irrigation engineering and more. 335 illustrations. Exercises, with answers. 784pp. 6⅛ x 9¼.
0-486-65675-6

THEORY OF VISCOELASTICITY (Second Edition), Richard M. Christensen. Complete consistent description of the linear theory of the viscoelastic behavior of materials. Problem-solving techniques discussed. 1982 edition. 29 figures. xiv+364pp. 6⅛ x 9¼. 0-486-42880-X

MECHANICS, J. P. Den Hartog. A classic introductory text or refresher. Hundreds of applications and design problems illuminate fundamentals of trusses, loaded beams and cables, etc. 334 answered problems. 462pp. 5⅜ x 8½. 0-486-60754-2

MECHANICAL VIBRATIONS, J. P. Den Hartog. Classic textbook offers lucid explanations and illustrative models, applying theories of vibrations to a variety of practical industrial engineering problems. Numerous figures. 233 problems, solutions. Appendix. Index. Preface. 436pp. 5⅜ x 8½. 0-486-64785-4

STRENGTH OF MATERIALS, J. P. Den Hartog. Full, clear treatment of basic material (tension, torsion, bending, etc.) plus advanced material on engineering methods, applications. 350 answered problems. 323pp. 5⅜ x 8½. 0-486-60755-0

A HISTORY OF MECHANICS, René Dugas. Monumental study of mechanical principles from antiquity to quantum mechanics. Contributions of ancient Greeks, Galileo, Leonardo, Kepler, Lagrange, many others. 671pp. 5⅜ x 8½. 0-486-65632-2

STABILITY THEORY AND ITS APPLICATIONS TO STRUCTURAL MECHANICS, Clive L. Dym. Self-contained text focuses on Koiter postbuckling analyses, with mathematical notions of stability of motion. Basing minimum energy principles for static stability upon dynamic concepts of stability of motion, it develops asymptotic buckling and postbuckling analyses from potential energy considerations, with applications to columns, plates, and arches. 1974 ed. 208pp. 5⅜ x 8½.
0-486-42541-X

METAL FATIGUE, N. E. Frost, K. J. Marsh, and L. P. Pook. Definitive, clearly written, and well-illustrated volume addresses all aspects of the subject, from the historical development of understanding metal fatigue to vital concepts of the cyclic stress that causes a crack to grow. Includes 7 appendixes. 544pp. 5⅜ x 8½. 0-486-40927-9

ROCKETS, Robert Goddard. Two of the most significant publications in the history of rocketry and jet propulsion: "A Method of Reaching Extreme Altitudes" (1919) and "Liquid Propellant Rocket Development" (1936). 128pp. 5⅜ x 8½.   0-486-42537-1

STATISTICAL MECHANICS: PRINCIPLES AND APPLICATIONS, Terrell L. Hill. Standard text covers fundamentals of statistical mechanics, applications to fluctuation theory, imperfect gases, distribution functions, more. 448pp. 5⅜ x 8½.

0-486-65390-0

ENGINEERING AND TECHNOLOGY 1650–1750: ILLUSTRATIONS AND TEXTS FROM ORIGINAL SOURCES, Martin Jensen. Highly readable text with more than 200 contemporary drawings and detailed engravings of engineering projects dealing with surveying, leveling, materials, hand tools, lifting equipment, transport and erection, piling, bailing, water supply, hydraulic engineering, and more. Among the specific projects outlined-transporting a 50-ton stone to the Louvre, erecting an obelisk, building timber locks, and dredging canals. 207pp. 8⅜ x 11¼.

0-486-42232-1

THE VARIATIONAL PRINCIPLES OF MECHANICS, Cornelius Lanczos. Graduate level coverage of calculus of variations, equations of motion, relativistic mechanics, more. First inexpensive paperbound edition of classic treatise. Index. Bibliography. 418pp. 5⅜ x 8½.   0-486-65067-7

PROTECTION OF ELECTRONIC CIRCUITS FROM OVERVOLTAGES, Ronald B. Standler. Five-part treatment presents practical rules and strategies for circuits designed to protect electronic systems from damage by transient overvoltages. 1989 ed. xxiv+434pp. 6⅛ x 9¼.   0-486-42552-5

ROTARY WING AERODYNAMICS, W. Z. Stepniewski. Clear, concise text covers aerodynamic phenomena of the rotor and offers guidelines for helicopter performance evaluation. Originally prepared for NASA. 537 figures. 640pp. 6⅛ x 9¼.

0-486-64647-5

INTRODUCTION TO SPACE DYNAMICS, William Tyrrell Thomson. Comprehensive, classic introduction to space-flight engineering for advanced undergraduate and graduate students. Includes vector algebra, kinematics, transformation of coordinates. Bibliography. Index. 352pp. 5⅜ x 8½.   0-486-65113-4

HISTORY OF STRENGTH OF MATERIALS, Stephen P. Timoshenko. Excellent historical survey of the strength of materials with many references to the theories of elasticity and structure. 245 figures. 452pp. 5⅜ x 8½.   0-486-61187-6

ANALYTICAL FRACTURE MECHANICS, David J. Unger. Self-contained text supplements standard fracture mechanics texts by focusing on analytical methods for determining crack-tip stress and strain fields. 336pp. 6⅛ x 9¼.   0-486-41737-9

STATISTICAL MECHANICS OF ELASTICITY, J. H. Weiner. Advanced, self-contained treatment illustrates general principles and elastic behavior of solids. Part 1, based on classical mechanics, studies thermoelastic behavior of crystalline and polymeric solids. Part 2, based on quantum mechanics, focuses on interatomic force laws, behavior of solids, and thermally activated processes. For students of physics and chemistry and for polymer physicists. 1983 ed. 96 figures. 496pp. 5⅜ x 8½.

0-486-42260-7

Mathematics

FUNCTIONAL ANALYSIS (Second Corrected Edition), George Bachman and Lawrence Narici. Excellent treatment of subject geared toward students with background in linear algebra, advanced calculus, physics and engineering. Text covers introduction to inner-product spaces, normed, metric spaces, and topological spaces; complete orthonormal sets, the Hahn-Banach Theorem and its consequences, and many other related subjects. 1966 ed. 544pp. 6⅛ x 9¼. 0-486-40251-7

ASYMPTOTIC EXPANSIONS OF INTEGRALS, Norman Bleistein & Richard A. Handelsman. Best introduction to important field with applications in a variety of scientific disciplines. New preface. Problems. Diagrams. Tables. Bibliography. Index. 448pp. 5⅜ x 8½. 0-486-65082-0

VECTOR AND TENSOR ANALYSIS WITH APPLICATIONS, A. I. Borisenko and I. E. Tarapov. Concise introduction. Worked-out problems, solutions, exercises. 257pp. 5⅝ x 8¼. 0-486-63833-2

AN INTRODUCTION TO ORDINARY DIFFERENTIAL EQUATIONS, Earl A. Coddington. A thorough and systematic first course in elementary differential equations for undergraduates in mathematics and science, with many exercises and problems (with answers). Index. 304pp. 5⅜ x 8½. 0-486-65942-9

FOURIER SERIES AND ORTHOGONAL FUNCTIONS, Harry F. Davis. An incisive text combining theory and practical example to introduce Fourier series, orthogonal functions and applications of the Fourier method to boundary-value problems. 570 exercises. Answers and notes. 416pp. 5⅜ x 8½. 0-486-65973-9

COMPUTABILITY AND UNSOLVABILITY, Martin Davis. Classic graduate-level introduction to theory of computability, usually referred to as theory of recurrent functions. New preface and appendix. 288pp. 5⅜ x 8½. 0-486-61471-9

ASYMPTOTIC METHODS IN ANALYSIS, N. G. de Bruijn. An inexpensive, comprehensive guide to asymptotic methods–the pioneering work that teaches by explaining worked examples in detail. Index. 224pp. 5⅜ x 8½ 0-486-64221-6

APPLIED COMPLEX VARIABLES, John W. Dettman. Step-by-step coverage of fundamentals of analytic function theory–plus lucid exposition of five important applications: Potential Theory; Ordinary Differential Equations; Fourier Transforms; Laplace Transforms; Asymptotic Expansions. 66 figures. Exercises at chapter ends. 512pp. 5⅜ x 8½. 0-486-64670-X

INTRODUCTION TO LINEAR ALGEBRA AND DIFFERENTIAL EQUATIONS, John W. Dettman. Excellent text covers complex numbers, determinants, orthonormal bases, Laplace transforms, much more. Exercises with solutions. Undergraduate level. 416pp. 5⅜ x 8½. 0-486-65191-6

RIEMANN'S ZETA FUNCTION, H. M. Edwards. Superb, high-level study of landmark 1859 publication entitled "On the Number of Primes Less Than a Given Magnitude" traces developments in mathematical theory that it inspired. xiv+315pp. 5⅜ x 8½. 0-486-41740-9

CALCULUS OF VARIATIONS WITH APPLICATIONS, George M. Ewing. Applications-oriented introduction to variational theory develops insight and promotes understanding of specialized books, research papers. Suitable for advanced undergraduate/graduate students as primary, supplementary text. 352pp. 5⅜ x 8½. 0-486-64856-7

COMPLEX VARIABLES, Francis J. Flanigan. Unusual approach, delaying complex algebra till harmonic functions have been analyzed from real variable viewpoint. Includes problems with answers. 364pp. 5⅜ x 8½. 0-486-61388-7

AN INTRODUCTION TO THE CALCULUS OF VARIATIONS, Charles Fox. Graduate-level text covers variations of an integral, isoperimetrical problems, least action, special relativity, approximations, more. References. 279pp. 5⅜ x 8½. 0-486-65499-0

COUNTEREXAMPLES IN ANALYSIS, Bernard R. Gelbaum and John M. H. Olmsted. These counterexamples deal mostly with the part of analysis known as "real variables." The first half covers the real number system, and the second half encompasses higher dimensions. 1962 edition. xxiv+198pp. 5⅜ x 8½. 0-486-42875-3

CATASTROPHE THEORY FOR SCIENTISTS AND ENGINEERS, Robert Gilmore. Advanced-level treatment describes mathematics of theory grounded in the work of Poincaré, R. Thom, other mathematicians. Also important applications to problems in mathematics, physics, chemistry and engineering. 1981 edition. References. 28 tables. 397 black-and-white illustrations. xvii + 666pp. 6⅛ x 9¼. 0-486-67539-4

INTRODUCTION TO DIFFERENCE EQUATIONS, Samuel Goldberg. Exceptionally clear exposition of important discipline with applications to sociology, psychology, economics. Many illustrative examples; over 250 problems. 260pp. 5⅜ x 8½. 0-486-65084-7

NUMERICAL METHODS FOR SCIENTISTS AND ENGINEERS, Richard Hamming. Classic text stresses frequency approach in coverage of algorithms, polynomial approximation, Fourier approximation, exponential approximation, other topics. Revised and enlarged 2nd edition. 721pp. 5⅜ x 8½. 0-486-65241-6

INTRODUCTION TO NUMERICAL ANALYSIS (2nd Edition), F. B. Hildebrand. Classic, fundamental treatment covers computation, approximation, interpolation, numerical differentiation and integration, other topics. 150 new problems. 669pp. 5⅜ x 8½. 0-486-65363-3

THREE PEARLS OF NUMBER THEORY, A. Y. Khinchin. Three compelling puzzles require proof of a basic law governing the world of numbers. Challenges concern van der Waerden's theorem, the Landau-Schnirelmann hypothesis and Mann's theorem, and a solution to Waring's problem. Solutions included. 64pp. 5⅜ x 8½. 0-486-40026-3

THE PHILOSOPHY OF MATHEMATICS: AN INTRODUCTORY ESSAY, Stephan Körner. Surveys the views of Plato, Aristotle, Leibniz & Kant concerning propositions and theories of applied and pure mathematics. Introduction. Two appendices. Index. 198pp. 5⅜ x 8½. 0-486-25048-2

INTRODUCTORY REAL ANALYSIS, A.N. Kolmogorov, S. V. Fomin. Translated by Richard A. Silverman. Self-contained, evenly paced introduction to real and functional analysis. Some 350 problems. 403pp. 5⅜ x 8½.   0-486-61226-0

APPLIED ANALYSIS, Cornelius Lanczos. Classic work on analysis and design of finite processes for approximating solution of analytical problems. Algebraic equations, matrices, harmonic analysis, quadrature methods, much more. 559pp. 5⅜ x 8½.
0-486-65656-X

AN INTRODUCTION TO ALGEBRAIC STRUCTURES, Joseph Landin. Superb self-contained text covers "abstract algebra": sets and numbers, theory of groups, theory of rings, much more. Numerous well-chosen examples, exercises. 247pp. 5⅜ x 8½.
0-486-65940-2

QUALITATIVE THEORY OF DIFFERENTIAL EQUATIONS, V. V. Nemytskii and V.V. Stepanov. Classic graduate-level text by two prominent Soviet mathematicians covers classical differential equations as well as topological dynamics and ergodic theory. Bibliographies. 523pp. 5⅜ x 8½.   0-486-65954-2

THEORY OF MATRICES, Sam Perlis. Outstanding text covering rank, nonsingularity and inverses in connection with the development of canonical matrices under the relation of equivalence, and without the intervention of determinants. Includes exercises. 237pp. 5⅜ x 8½.   0-486-66810-X

INTRODUCTION TO ANALYSIS, Maxwell Rosenlicht. Unusually clear, accessible coverage of set theory, real number system, metric spaces, continuous functions, Riemann integration, multiple integrals, more. Wide range of problems. Undergraduate level. Bibliography. 254pp. 5⅜ x 8½.   0-486-65038-3

MODERN NONLINEAR EQUATIONS, Thomas L. Saaty. Emphasizes practical solution of problems; covers seven types of equations. ". . . a welcome contribution to the existing literature...."–*Math Reviews.* 490pp. 5⅜ x 8½.   0-486-64232-1

MATRICES AND LINEAR ALGEBRA, Hans Schneider and George Phillip Barker. Basic textbook covers theory of matrices and its applications to systems of linear equations and related topics such as determinants, eigenvalues and differential equations. Numerous exercises. 432pp. 5⅜ x 8½.   0-486-66014-1

LINEAR ALGEBRA, Georgi E. Shilov. Determinants, linear spaces, matrix algebras, similar topics. For advanced undergraduates, graduates. Silverman translation. 387pp. 5⅜ x 8½.   0-486-63518-X

ELEMENTS OF REAL ANALYSIS, David A. Sprecher. Classic text covers fundamental concepts, real number system, point sets, functions of a real variable, Fourier series, much more. Over 500 exercises. 352pp. 5⅜ x 8½.   0-486-65385-4

SET THEORY AND LOGIC, Robert R. Stoll. Lucid introduction to unified theory of mathematical concepts. Set theory and logic seen as tools for conceptual understanding of real number system. 496pp. 5⅜ x 8¼.   0-486-63829-4

TENSOR CALCULUS, J.L. Synge and A. Schild. Widely used introductory text covers spaces and tensors, basic operations in Riemannian space, non-Riemannian spaces, etc. 324pp. 5⅜ x 8¼. 0-486-63612-7

ORDINARY DIFFERENTIAL EQUATIONS, Morris Tenenbaum and Harry Pollard. Exhaustive survey of ordinary differential equations for undergraduates in mathematics, engineering, science. Thorough analysis of theorems. Diagrams. Bibliography. Index. 818pp. 5⅜ x 8½. 0-486-64940-7

INTEGRAL EQUATIONS, F. G. Tricomi. Authoritative, well-written treatment of extremely useful mathematical tool with wide applications. Volterra Equations, Fredholm Equations, much more. Advanced undergraduate to graduate level. Exercises. Bibliography. 238pp. 5⅜ x 8½. 0-486-64828-1

FOURIER SERIES, Georgi P. Tolstov. Translated by Richard A. Silverman. A valuable addition to the literature on the subject, moving clearly from subject to subject and theorem to theorem. 107 problems, answers. 336pp. 5⅜ x 8½. 0-486-63317-9

INTRODUCTION TO MATHEMATICAL THINKING, Friedrich Waismann. Examinations of arithmetic, geometry, and theory of integers; rational and natural numbers; complete induction; limit and point of accumulation; remarkable curves; complex and hypercomplex numbers, more. 1959 ed. 27 figures. xii+260pp. 5⅜ x 8½. 0-486-63317-9

POPULAR LECTURES ON MATHEMATICAL LOGIC, Hao Wang. Noted logician's lucid treatment of historical developments, set theory, model theory, recursion theory and constructivism, proof theory, more. 3 appendixes. Bibliography. 1981 edition. ix + 283pp. 5⅜ x 8½. 0-486-67632-3

CALCULUS OF VARIATIONS, Robert Weinstock. Basic introduction covering isoperimetric problems, theory of elasticity, quantum mechanics, electrostatics, etc. Exercises throughout. 326pp. 5⅜ x 8½. 0-486-63069-2

THE CONTINUUM: A CRITICAL EXAMINATION OF THE FOUNDATION OF ANALYSIS, Hermann Weyl. Classic of 20th-century foundational research deals with the conceptual problem posed by the continuum. 156pp. 5⅜ x 8½. 0-486-67982-9

CHALLENGING MATHEMATICAL PROBLEMS WITH ELEMENTARY SOLUTIONS, A. M. Yaglom and I. M. Yaglom. Over 170 challenging problems on probability theory, combinatorial analysis, points and lines, topology, convex polygons, many other topics. Solutions. Total of 445pp. 5⅜ x 8½. Two-vol. set.
Vol. I: 0-486-65536-9 Vol. II: 0-486-65537-7

INTRODUCTION TO PARTIAL DIFFERENTIAL EQUATIONS WITH APPLICATIONS, E. C. Zachmanoglou and Dale W. Thoe. Essentials of partial differential equations applied to common problems in engineering and the physical sciences. Problems and answers. 416pp. 5⅜ x 8½. 0-486-65251-3

THE THEORY OF GROUPS, Hans J. Zassenhaus. Well-written graduate-level text acquaints reader with group-theoretic methods and demonstrates their usefulness in mathematics. Axioms, the calculus of complexes, homomorphic mapping, *p*-group theory, more. 276pp. 5⅜ x 8½. 0-486-40922-8

Math–Decision Theory, Statistics, Probability

ELEMENTARY DECISION THEORY, Herman Chernoff and Lincoln E. Moses. Clear introduction to statistics and statistical theory covers data processing, probability and random variables, testing hypotheses, much more. Exercises. 364pp. 5⅜ x 8½.   0-486-65218-1

STATISTICS MANUAL, Edwin L. Crow et al. Comprehensive, practical collection of classical and modern methods prepared by U.S. Naval Ordnance Test Station. Stress on use. Basics of statistics assumed. 288pp. 5⅜ x 8½.   0-486-60599-X

SOME THEORY OF SAMPLING, William Edwards Deming. Analysis of the problems, theory and design of sampling techniques for social scientists, industrial managers and others who find statistics important at work. 61 tables. 90 figures. xvii +602pp. 5⅜ x 8½.   0-486-64684-X

LINEAR PROGRAMMING AND ECONOMIC ANALYSIS, Robert Dorfman, Paul A. Samuelson and Robert M. Solow. First comprehensive treatment of linear programming in standard economic analysis. Game theory, modern welfare economics, Leontief input-output, more. 525pp. 5⅜ x 8½.   0-486-65491-5

PROBABILITY: AN INTRODUCTION, Samuel Goldberg. Excellent basic text covers set theory, probability theory for finite sample spaces, binomial theorem, much more. 360 problems. Bibliographies. 322pp. 5⅜ x 8½.   0-486-65252-1

GAMES AND DECISIONS: INTRODUCTION AND CRITICAL SURVEY, R. Duncan Luce and Howard Raiffa. Superb nontechnical introduction to game theory, primarily applied to social sciences. Utility theory, zero-sum games, n-person games, decision-making, much more. Bibliography. 509pp. 5⅜ x 8½. 0-486-65943-7

INTRODUCTION TO THE THEORY OF GAMES, J. C. C. McKinsey. This comprehensive overview of the mathematical theory of games illustrates applications to situations involving conflicts of interest, including economic, social, political, and military contexts. Appropriate for advanced undergraduate and graduate courses; advanced calculus a prerequisite. 1952 ed. x+372pp. 5⅜ x 8½.   0-486-42811-7

FIFTY CHALLENGING PROBLEMS IN PROBABILITY WITH SOLUTIONS, Frederick Mosteller. Remarkable puzzlers, graded in difficulty, illustrate elementary and advanced aspects of probability. Detailed solutions. 88pp. 5⅜ x 8½.   65355-2

PROBABILITY THEORY: A CONCISE COURSE, Y. A. Rozanov. Highly readable, self-contained introduction covers combination of events, dependent events, Bernoulli trials, etc. 148pp. 5⅜ x 8¼.   0-486-63544-9

STATISTICAL METHOD FROM THE VIEWPOINT OF QUALITY CONTROL, Walter A. Shewhart. Important text explains regulation of variables, uses of statistical control to achieve quality control in industry, agriculture, other areas. 192pp. 5⅜ x 8½.   0-486-65232-7

Math–Geometry and Topology

ELEMENTARY CONCEPTS OF TOPOLOGY, Paul Alexandroff. Elegant, intuitive approach to topology from set-theoretic topology to Betti groups; how concepts of topology are useful in math and physics. 25 figures. 57pp. 5⅜ x 8½.   0-486-60747-X

COMBINATORIAL TOPOLOGY, P. S. Alexandrov. Clearly written, well-organized, three-part text begins by dealing with certain classic problems without using the formal techniques of homology theory and advances to the central concept, the Betti groups. Numerous detailed examples. 654pp. 5¾ x 8¼.   0-486-40179-0

EXPERIMENTS IN TOPOLOGY, Stephen Barr. Classic, lively explanation of one of the byways of mathematics. Klein bottles, Moebius strips, projective planes, map coloring, problem of the Koenigsberg bridges, much more, described with clarity and wit. 43 figures. 210pp. 5⅜ x 8½.   0-486-25933-1

THE GEOMETRY OF RENÉ DESCARTES, René Descartes. The great work founded analytical geometry. Original French text, Descartes's own diagrams, together with definitive Smith-Latham translation. 244pp. 5⅜ x 8½.   0-486-60068-8

EUCLIDEAN GEOMETRY AND TRANSFORMATIONS, Clayton W. Dodge. This introduction to Euclidean geometry emphasizes transformations, particularly isometries and similarities. Suitable for undergraduate courses, it includes numerous examples, many with detailed answers. 1972 ed. viii+296pp. 6⅛ x 9¼. 0-486-43476-1

PRACTICAL CONIC SECTIONS: THE GEOMETRIC PROPERTIES OF ELLIPSES, PARABOLAS AND HYPERBOLAS, J. W. Downs. This text shows how to create ellipses, parabolas, and hyperbolas. It also presents historical background on their ancient origins and describes the reflective properties and roles of curves in design applications. 1993 ed. 98 figures. xii+100pp. 6½ x 9¼.   0-486-42876-1

THE THIRTEEN BOOKS OF EUCLID'S ELEMENTS, translated with introduction and commentary by Sir Thomas L. Heath. Definitive edition. Textual and linguistic notes, mathematical analysis. 2,500 years of critical commentary. Unabridged. 1,414pp. 5⅜ x 8½. Three-vol. set.
Vol. I: 0-486-60088-2   Vol. II: 0-486-60089-0   Vol. III: 0-486-60090-4

SPACE AND GEOMETRY: IN THE LIGHT OF PHYSIOLOGICAL, PSYCHOLOGICAL AND PHYSICAL INQUIRY, Ernst Mach. Three essays by an eminent philosopher and scientist explore the nature, origin, and development of our concepts of space, with a distinctness and precision suitable for undergraduate students and other readers. 1906 ed. vi+148pp. 5⅜ x 8½.   0-486-43909-7

GEOMETRY OF COMPLEX NUMBERS, Hans Schwerdtfeger. Illuminating, widely praised book on analytic geometry of circles, the Moebius transformation, and two-dimensional non-Euclidean geometries. 200pp. 5⅜ x 8¼.   0-486-63830-8

DIFFERENTIAL GEOMETRY, Heinrich W. Guggenheimer. Local differential geometry as an application of advanced calculus and linear algebra. Curvature, transformation groups, surfaces, more. Exercises. 62 figures. 378pp. 5⅜ x 8½.   0-486-63433-7

History of Math

THE WORKS OF ARCHIMEDES, Archimedes (T. L. Heath, ed.). Topics include the famous problems of the ratio of the areas of a cylinder and an inscribed sphere; the measurement of a circle; the properties of conoids, spheroids, and spirals; and the quadrature of the parabola. Informative introduction. clxxxvi+326pp. 5⅜ x 8½.
0-486-42084-1

A SHORT ACCOUNT OF THE HISTORY OF MATHEMATICS, W. W. Rouse Ball. One of clearest, most authoritative surveys from the Egyptians and Phoenicians through 19th-century figures such as Grassman, Galois, Riemann. Fourth edition. 522pp. 5⅜ x 8½. 0-486-20630-0

THE HISTORY OF THE CALCULUS AND ITS CONCEPTUAL DEVELOP-MENT, Carl B. Boyer. Origins in antiquity, medieval contributions, work of Newton, Leibniz, rigorous formulation. Treatment is verbal. 346pp. 5⅜ x 8½. 0-486-60509-4

THE HISTORICAL ROOTS OF ELEMENTARY MATHEMATICS, Lucas N. H. Bunt, Phillip S. Jones, and Jack D. Bedient. Fundamental underpinnings of modern arithmetic, algebra, geometry and number systems derived from ancient civilizations. 320pp. 5⅜ x 8½. 0-486-25563-8

A HISTORY OF MATHEMATICAL NOTATIONS, Florian Cajori. This classic study notes the first appearance of a mathematical symbol and its origin, the competition it encountered, its spread among writers in different countries, its rise to popularity, its eventual decline or ultimate survival. Original 1929 two-volume edition presented here in one volume. xxviii+820pp. 5⅜ x 8½. 0-486-67766-4

GAMES, GODS & GAMBLING: A HISTORY OF PROBABILITY AND STATISTICAL IDEAS, F. N. David. Episodes from the lives of Galileo, Fermat, Pascal, and others illustrate this fascinating account of the roots of mathematics. Features thought-provoking references to classics, archaeology, biography, poetry. 1962 edition. 304pp. 5⅜ x 8½. (Available in U.S. only.) 0-486-40023-9

OF MEN AND NUMBERS: THE STORY OF THE GREAT MATHEMATICIANS, Jane Muir. Fascinating accounts of the lives and accomplishments of history's greatest mathematical minds—Pythagoras, Descartes, Euler, Pascal, Cantor, many more. Anecdotal, illuminating. 30 diagrams. Bibliography. 256pp. 5⅜ x 8½. 0-486-28973-7

HISTORY OF MATHEMATICS, David E. Smith. Nontechnical survey from ancient Greece and Orient to late 19th century; evolution of arithmetic, geometry, trigonometry, calculating devices, algebra, the calculus. 362 illustrations. 1,355pp. 5⅜ x 8½. Two-vol. set. Vol. I: 0-486-20429-4 Vol. II: 0-486-20430-8

A CONCISE HISTORY OF MATHEMATICS, Dirk J. Struik. The best brief history of mathematics. Stresses origins and covers every major figure from ancient Near East to 19th century. 41 illustrations. 195pp. 5⅜ x 8½. 0-486-60255-9

Physics

OPTICAL RESONANCE AND TWO-LEVEL ATOMS, L. Allen and J. H. Eberly. Clear, comprehensive introduction to basic principles behind all quantum optical resonance phenomena. 53 illustrations. Preface. Index. 256pp. 5⅜ x 8½. 0-486-65533-4

QUANTUM THEORY, David Bohm. This advanced undergraduate-level text presents the quantum theory in terms of qualitative and imaginative concepts, followed by specific applications worked out in mathematical detail. Preface. Index. 655pp. 5⅜ x 8½. 0-486-65969-0

ATOMIC PHYSICS (8th EDITION), Max Born. Nobel laureate's lucid treatment of kinetic theory of gases, elementary particles, nuclear atom, wave-corpuscles, atomic structure and spectral lines, much more. Over 40 appendices, bibliography. 495pp. 5⅜ x 8½. 0-486-65984-4

A SOPHISTICATE'S PRIMER OF RELATIVITY, P. W. Bridgman. Geared toward readers already acquainted with special relativity, this book transcends the view of theory as a working tool to answer natural questions: What is a frame of reference? What is a "law of nature"? What is the role of the "observer"? Extensive treatment, written in terms accessible to those without a scientific background. 1983 ed. xlviii+172pp. 5⅜ x 8½. 0-486-42549-5

AN INTRODUCTION TO HAMILTONIAN OPTICS, H. A. Buchdahl. Detailed account of the Hamiltonian treatment of aberration theory in geometrical optics. Many classes of optical systems defined in terms of the symmetries they possess. Problems with detailed solutions. 1970 edition. xv + 360pp. 5⅜ x 8½. 0-486-67597-1

PRIMER OF QUANTUM MECHANICS, Marvin Chester. Introductory text examines the classical quantum bead on a track: its state and representations; operator eigenvalues; harmonic oscillator and bound bead in a symmetric force field; and bead in a spherical shell. Other topics include spin, matrices, and the structure of quantum mechanics; the simplest atom; indistinguishable particles; and stationary-state perturbation theory. 1992 ed. xiv+314pp. 6⅛ x 9¼. 0-486-42878-8

LECTURES ON QUANTUM MECHANICS, Paul A. M. Dirac. Four concise, brilliant lectures on mathematical methods in quantum mechanics from Nobel Prize-winning quantum pioneer build on idea of visualizing quantum theory through the use of classical mechanics. 96pp. 5⅜ x 8½. 0-486-41713-1

THIRTY YEARS THAT SHOOK PHYSICS: THE STORY OF QUANTUM THEORY, George Gamow. Lucid, accessible introduction to influential theory of energy and matter. Careful explanations of Dirac's anti-particles, Bohr's model of the atom, much more. 12 plates. Numerous drawings. 240pp. 5⅜ x 8½. 0-486-24895-X

ELECTRONIC STRUCTURE AND THE PROPERTIES OF SOLIDS: THE PHYSICS OF THE CHEMICAL BOND, Walter A. Harrison. Innovative text offers basic understanding of the electronic structure of covalent and ionic solids, simple metals, transition metals and their compounds. Problems. 1980 edition. 582pp. 6⅛ x 9¼. 0-486-66021-4

HYDRODYNAMIC AND HYDROMAGNETIC STABILITY, S. Chandrasekhar. Lucid examination of the Rayleigh-Benard problem; clear coverage of the theory of instabilities causing convection. 704pp. 5⅜ x 8¼.   0-486-64071-X

INVESTIGATIONS ON THE THEORY OF THE BROWNIAN MOVEMENT, Albert Einstein. Five papers (1905–8) investigating dynamics of Brownian motion and evolving elementary theory. Notes by R. Fürth. 122pp. 5⅜ x 8½. 0-486-60304-0

THE PHYSICS OF WAVES, William C. Elmore and Mark A. Heald. Unique overview of classical wave theory. Acoustics, optics, electromagnetic radiation, more. Ideal as classroom text or for self-study. Problems. 477pp. 5⅜ x 8½.   0-486-64926-1

GRAVITY, George Gamow. Distinguished physicist and teacher takes reader-friendly look at three scientists whose work unlocked many of the mysteries behind the laws of physics: Galileo, Newton, and Einstein. Most of the book focuses on Newton's ideas, with a concluding chapter on post-Einsteinian speculations concerning the relationship between gravity and other physical phenomena. 160pp. 5⅜ x 8½.
0-486-42563-0

PHYSICAL PRINCIPLES OF THE QUANTUM THEORY, Werner Heisenberg. Nobel Laureate discusses quantum theory, uncertainty, wave mechanics, work of Dirac, Schroedinger, Compton, Wilson, Einstein, etc. 184pp. 5⅜ x 8½. 0-486-60113-7

ATOMIC SPECTRA AND ATOMIC STRUCTURE, Gerhard Herzberg. One of best introductions; especially for specialist in other fields. Treatment is physical rather than mathematical. 80 illustrations. 257pp. 5⅜ x 8½.   0-486-60115-3

AN INTRODUCTION TO STATISTICAL THERMODYNAMICS, Terrell L. Hill. Excellent basic text offers wide-ranging coverage of quantum statistical mechanics, systems of interacting molecules, quantum statistics, more. 523pp. 5⅜ x 8½.
0-486-65242-4

THEORETICAL PHYSICS, Georg Joos, with Ira M. Freeman. Classic overview covers essential math, mechanics, electromagnetic theory, thermodynamics, quantum mechanics, nuclear physics, other topics. First paperback edition. xxiii + 885pp. 5⅜ x 8½.   0-486-65227-0

PROBLEMS AND SOLUTIONS IN QUANTUM CHEMISTRY AND PHYSICS, Charles S. Johnson, Jr. and Lee G. Pedersen. Unusually varied problems, detailed solutions in coverage of quantum mechanics, wave mechanics, angular momentum, molecular spectroscopy, more. 280 problems plus 139 supplementary exercises. 430pp. 6½ x 9¼.   0-486-65236-X

THEORETICAL SOLID STATE PHYSICS, Vol. 1: Perfect Lattices in Equilibrium; Vol. II: Non-Equilibrium and Disorder, William Jones and Norman H. March. Monumental reference work covers fundamental theory of equilibrium properties of perfect crystalline solids, non-equilibrium properties, defects and disordered systems. Appendices. Problems. Preface. Diagrams. Index. Bibliography. Total of 1,301pp. 5⅜ x 8½. Two volumes.   Vol. I: 0-486-65015-4   Vol. II: 0-486-65016-2

WHAT IS RELATIVITY? L. D. Landau and G. B. Rumer. Written by a Nobel Prize physicist and his distinguished colleague, this compelling book explains the special theory of relativity to readers with no scientific background, using such familiar objects as trains, rulers, and clocks. 1960 ed. vi+72pp. 5⅜ x 8½.   0-486-42806-0

CATALOG OF DOVER BOOKS

A TREATISE ON ELECTRICITY AND MAGNETISM, James Clerk Maxwell. Important foundation work of modern physics. Brings to final form Maxwell's theory of electromagnetism and rigorously derives his general equations of field theory. 1,084pp. 5⅜ x 8½. Two-vol. set. Vol. I: 0-486-60636-8 Vol. II: 0-486-60637-6

QUANTUM MECHANICS: PRINCIPLES AND FORMALISM, Roy McWeeny. Graduate student-oriented volume develops subject as fundamental discipline, opening with review of origins of Schrödinger's equations and vector spaces. Focusing on main principles of quantum mechanics and their immediate consequences, it concludes with final generalizations covering alternative "languages" or representations. 1972 ed. 15 figures. xi+155pp. 5⅜ x 8½. 0-486-42829-X

INTRODUCTION TO QUANTUM MECHANICS With Applications to Chemistry, Linus Pauling & E. Bright Wilson, Jr. Classic undergraduate text by Nobel Prize winner applies quantum mechanics to chemical and physical problems. Numerous tables and figures enhance the text. Chapter bibliographies. Appendices. Index. 468pp. 5⅜ x 8½. 0-486-64871-0

METHODS OF THERMODYNAMICS, Howard Reiss. Outstanding text focuses on physical technique of thermodynamics, typical problem areas of understanding, and significance and use of thermodynamic potential. 1965 edition. 238pp. 5⅜ x 8½. 0-486-69445-3

THE ELECTROMAGNETIC FIELD, Albert Shadowitz. Comprehensive undergraduate text covers basics of electric and magnetic fields, builds up to electromagnetic theory. Also related topics, including relativity. Over 900 problems. 768pp. 5⅝ x 8¼. 0-486-65660-8

GREAT EXPERIMENTS IN PHYSICS: FIRSTHAND ACCOUNTS FROM GALILEO TO EINSTEIN, Morris H. Shamos (ed.). 25 crucial discoveries: Newton's laws of motion, Chadwick's study of the neutron, Hertz on electromagnetic waves, more. Original accounts clearly annotated. 370pp. 5⅜ x 8½. 0-486-25346-5

EINSTEIN'S LEGACY, Julian Schwinger. A Nobel Laureate relates fascinating story of Einstein and development of relativity theory in well-illustrated, nontechnical volume. Subjects include meaning of time, paradoxes of space travel, gravity and its effect on light, non-Euclidean geometry and curving of space-time, impact of radio astronomy and space-age discoveries, and more. 189 b/w illustrations. xiv+250pp. 8⅜ x 9¼. 0-486-41974-6

STATISTICAL PHYSICS, Gregory H. Wannier. Classic text combines thermodynamics, statistical mechanics and kinetic theory in one unified presentation of thermal physics. Problems with solutions. Bibliography. 532pp. 5⅜ x 8½. 0-486-65401-X

Paperbound unless otherwise indicated. Available at your book dealer, online at **www.doverpublications.com**, or by writing to Dept. GI, Dover Publications, Inc., 31 East 2nd Street, Mineola, NY 11501. For current price information or for free catalogues (please indicate field of interest), write to Dover Publications or log on to **www.doverpublications.com** and see every Dover book in print. Dover publishes more than 500 books each year on science, elementary and advanced mathematics, biology, music, art, literary history, social sciences, and other areas.

$$(x^3 - 8)$$

$$(x + 2)(x - 2)(x - 2)$$

$$(x^2 - 4)(x + a)$$

$$(x^2 - 4)(x + a) = x^3 + ax^2 -$$

$$\begin{array}{r} x^2 +-2x + 4 \\ x-2\ \overline{\smash{\big)}\ x^3 - 8} \\ x^3 - 2x^2 \\ \hline -(8 + 2x^2) \\ +4x - 2x^2 \\ \hline 4x - 8 \end{array}$$